HOME REPAIR AND IMPROVEMENT

# ADVANCED MASONRY

OTHER PUBLICATIONS:

DO IT YOURSELF
The Time-Life Complete Gardener
Home Repair and Improvement
The Art of Woodworking
Fix It Yourself

COOKING
Weight Watchers® Smart Choice Recipe Collection
Great Taste/Low Fat
Williams-Sonoma Kitchen Library

HISTORY
The American Story
Voices of the Civil War
The American Indians
Lost Civilizations
Mysteries of the Unknown
Time Frame
The Civil War
Cultural Atlas

TIME-LIFE KIDS
Library of First Questions and Answers
A Child's First Library of Learning
I Love Math
Nature Company Discoveries
Understanding Science & Nature

SCIENCE/NATURE
Voyage Through the Universe

For information on and a full description
of any of the Time-Life Books series listed above,
please call 1-800-621-7026 or write:

Reader Information
Time-Life Customer Service
P.O. Box C-32068
Richmond Virginia 23261-2068

HOME REPAIR AND IMPROVEMENT

# ADVANCED MASONRY



BY THE EDITORS OF TIME-LIFE BOOKS, ALEXANDRIA, VIRGINIA

*The Consultants*

Robert S. Cook has taught concrete technology for several years. A concrete-finishing specialist with the Portland Cement Association for five years, he founded the Concrete Technician Program at L. H. Bates Vocational-Technical Institute in Tacoma, Washington.

Richard Day spent eight years with the Portland Cement Association as a writer and editor on subjects relating to concrete. Based in southern California, he has penned numerous articles for *Popular Science* and has authored several books on concrete and masonry.

Richard T. Kreh, Sr., has over 40 years of experience as a Master Mason. He has served as an instructor of masonry and building trades in Maryland, and has written articles for many magazines, including *Popular Science* and *Fine Homebuilding*. Mr. Kreh is also the author of three major textbooks on masonry.

# CONTENTS

# Mastering a Rugged Material

Beyond the ability to bind bricks and blocks with mortar, there are a number of skills that identify an accomplished mason. Some are aesthetic—for example, how to select the most appropriate material from the array of choices available. Others are more practical: how to cleave and shape stone, cut or drill into masonry structures, fasten heavy objects to them, and even, on occasion, to demolish them.



Cutting stone →

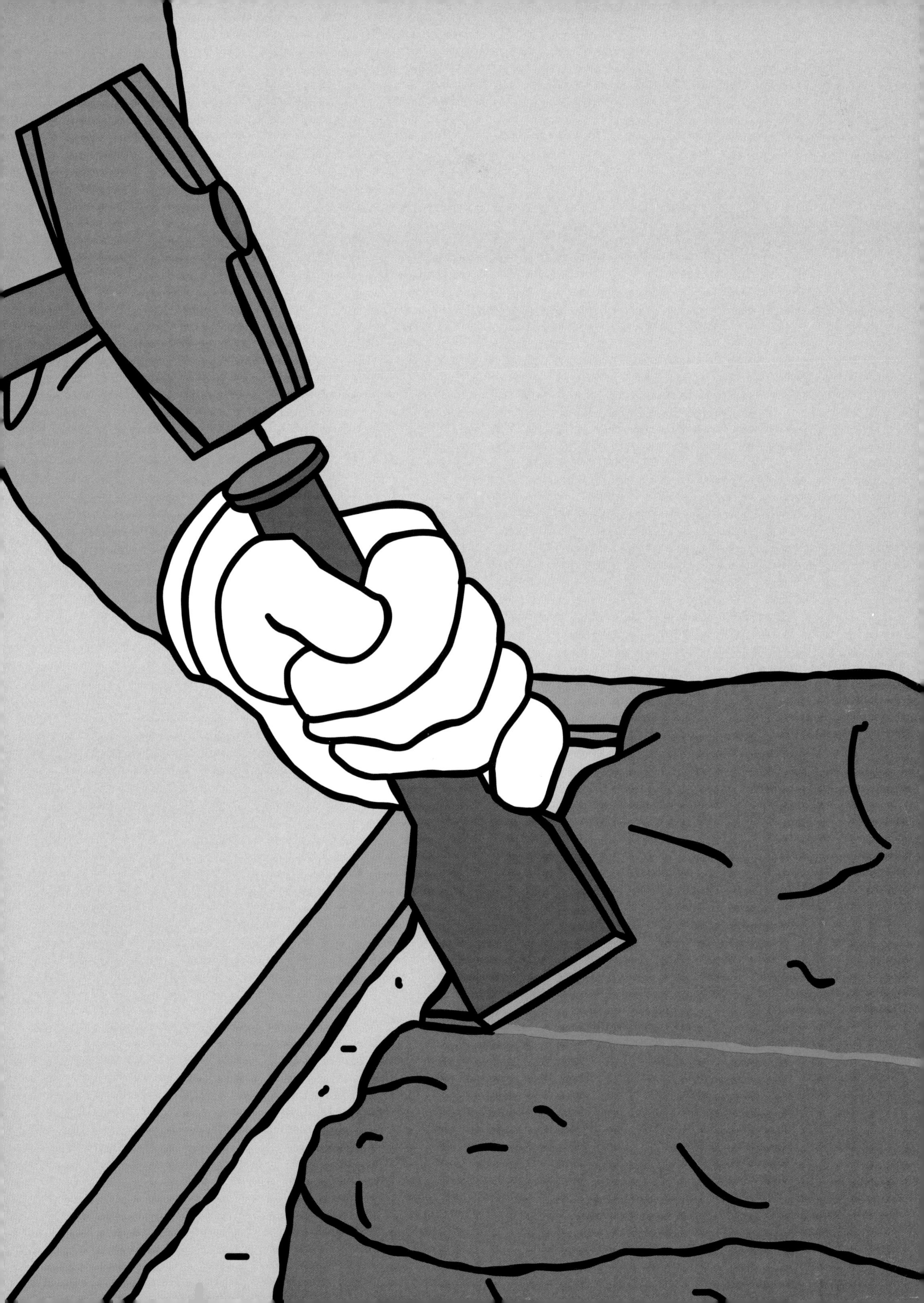

# A Selection of Bricks, Standard and Custom

Bricks are a versatile and durable building material, and a basic unit of the mason's craft. Most wall construction calls for two types of brick: the economical building brick and face brick *(below)*. Face bricks, available in many colors and surface textures, are used when a uniform appearance is called for. The surface finish may appear on only one of the long, narrow sides, called the face.

**Sizes:** A standard brick is nominally 8 inches by 4 inches by $2\frac{2}{3}$ inches—although its actual dimensions are slightly less, to allow for mortar joints. These nominal dimensions are all based on even fractions of the brick's length. Thus, a brick is half as wide as it is long, and a third as high, permitting bricks to be laid in a variety of patterns while maintaining unbroken horizontal joints. Larger or smaller bricks are sometimes used for aesthetic effect or to speed a job by using fewer bricks to fill a space. Specialty bricks are also available for trim applications *(opposite)*.

**Grades:** Bricks are available in three grades, based on their resistance to weathering. Those graded SW can withstand a high degree of frost action, as might occur in belowground foundations or retaining walls. Grade MW is used for less severe weather conditions, such as in an aboveground wall, where frost is common but the bricks are unlikely to be permeated with water. Grade NW applies only to building bricks and denotes minimal weather resistance. These units are used only for backup or interior walls. Face bricks are always rated either SW or MW.

**Ordering:** Buy face bricks for a project from one batch; the material and manufacturing procedures may change from one lot to another, resulting in unacceptable variations in color and texture. If you are ordering special bricks as part of the job, make sure they are from the same batch of clay as the standard ones in your order.

Bricks vary in water absorption from batch to batch: Those that are too dry may draw water from fresh mortar, preventing the formation of a sound joint. Test a brick from each lot by sprinkling a few drops of water on its broad surface, called the bed. If the water is completely absorbed in less than one minute, hose down the entire set of bricks. Then allow the surface water to be absorbed before laying the bricks.

The smallest number of standard bricks you can buy economically is a "strap"—100 pieces bound together by a metal band. Less expensive per brick is a cube, or 500 bricks. Have the straps or cubes delivered on wooden pallets; otherwise they may be unloaded into a jumbled pile, with possible breakage.

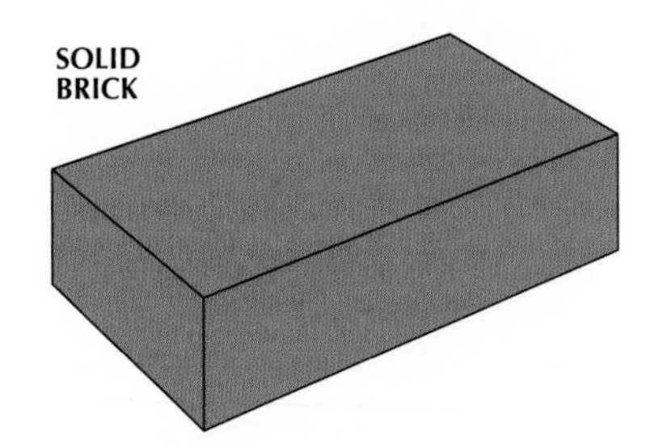


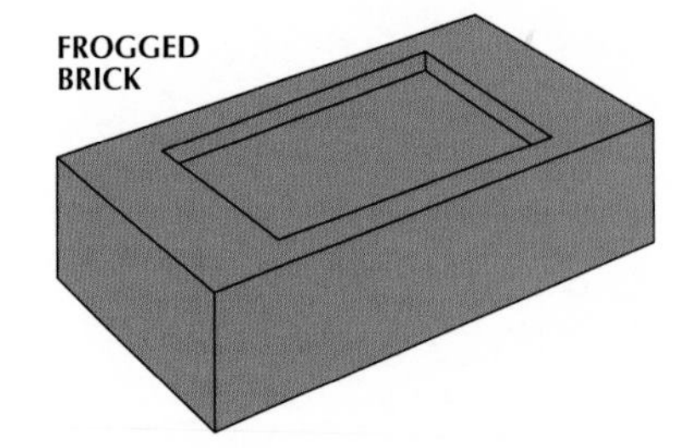


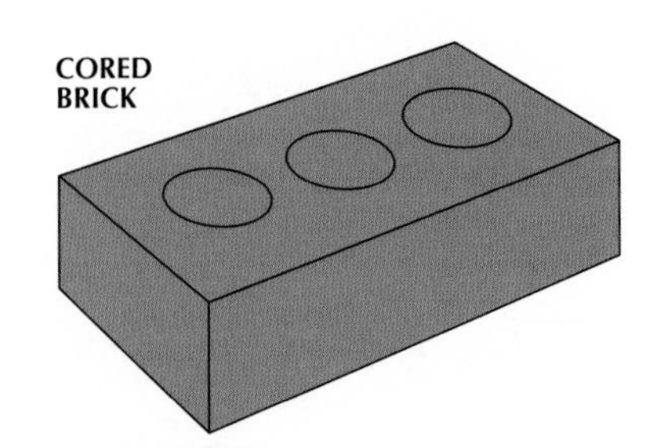


**Standard bricks.**
Building bricks and face bricks are sold in three forms. Solid bricks are flat on all sides. Frogged bricks—also solid—have a shallow depression called a frog on one of the two bed faces, making them lighter and allowing stronger mortar bonds. Cored bricks are made with holes cut all the way through them to reduce their weight.

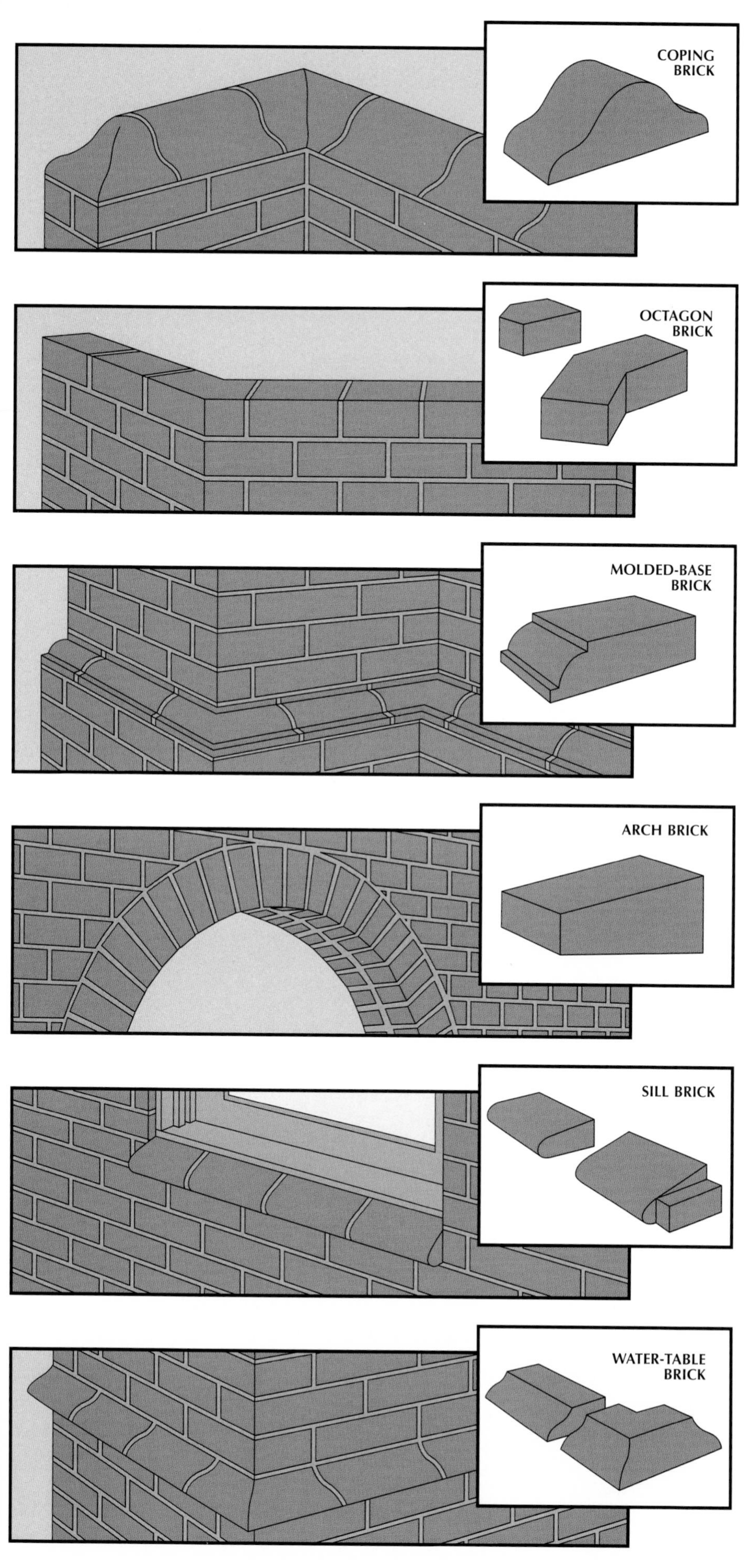


**Custom bricks.**
An ordinary brick wall can be dressed up with trim bricks, such as in the selection illustrated at left; these bricks often have to be specially ordered. Coping bricks provide an ornamental cap for a freestanding wall; ends and corners are finished with special matching units. Octagon bricks allow corners of 135 degrees—the angle of an octagon—and they come in two sizes to eliminate the need for cutting. Molded-base bricks finish the base of a wall or column, or fill the angle of a wall stepped back to reduce its thickness. Although arches can be made with standard bricks, special wedge-shaped arch bricks create a more regular appearance, allowing mortar joints of uniform thickness. Sill bricks create a slanting ledge under a window opening; special bricks for the end of the sill have tab-like extensions that are mortared into the adjoining brickwork. Water-table bricks are set in a wall 16 to 18 inches above the foundation to divert the vertical flow of water outward, away from the wall.

# Concrete Blocks for Plain and Fancy Walls

Concrete blocks are manufactured in fewer forms than brick, but they are a more economical building material.

**Standard and Lightweight Blocks:** Standard blocks are made with sand, gravel, and crushed stone, and weigh between 40 and 50 pounds. Lightweight blocks—25 to 35 pounds—contain pumice, expanded clay, shale, or slate. While both standard and lightweight blocks are suitable for applications where structural strength is required, such as load-bearing walls, standard blocks are preferred where sound reduction is required. Lightweight blocks have the advantage of being easier to lay. They are somewhat more expensive than standard blocks, but provide superior fire protection and insulation.

**Dimensions:** The nominal dimensions of a standard block, which allow for mortar joints, are 8 inches by 8 inches by 16 inches. Widths from 4 to 12 inches are also available to meet varying specifications for strength or wall thickness. Specially shaped blocks can be ordered in dimensions conforming to those of standard blocks.

Many of the specially shaped blocks are made for specific structural applications, such as tops and corners of walls. There also are ornamental blocks that relieve the visual monotony of a flat wall with texture, highlights, and shadows *(opposite)*.

**Ordering and Storing:** Concrete blocks are usually delivered on wooden pallets. Unless otherwise specified, each pallet usually contains a mixture of stretcher and corner blocks in the proportions required for most ordinary structures. Store the blocks on pallets and cover them with plastic. Never hose down blocks before or during construction, since a wall built with damp blocks may crack as it dries.

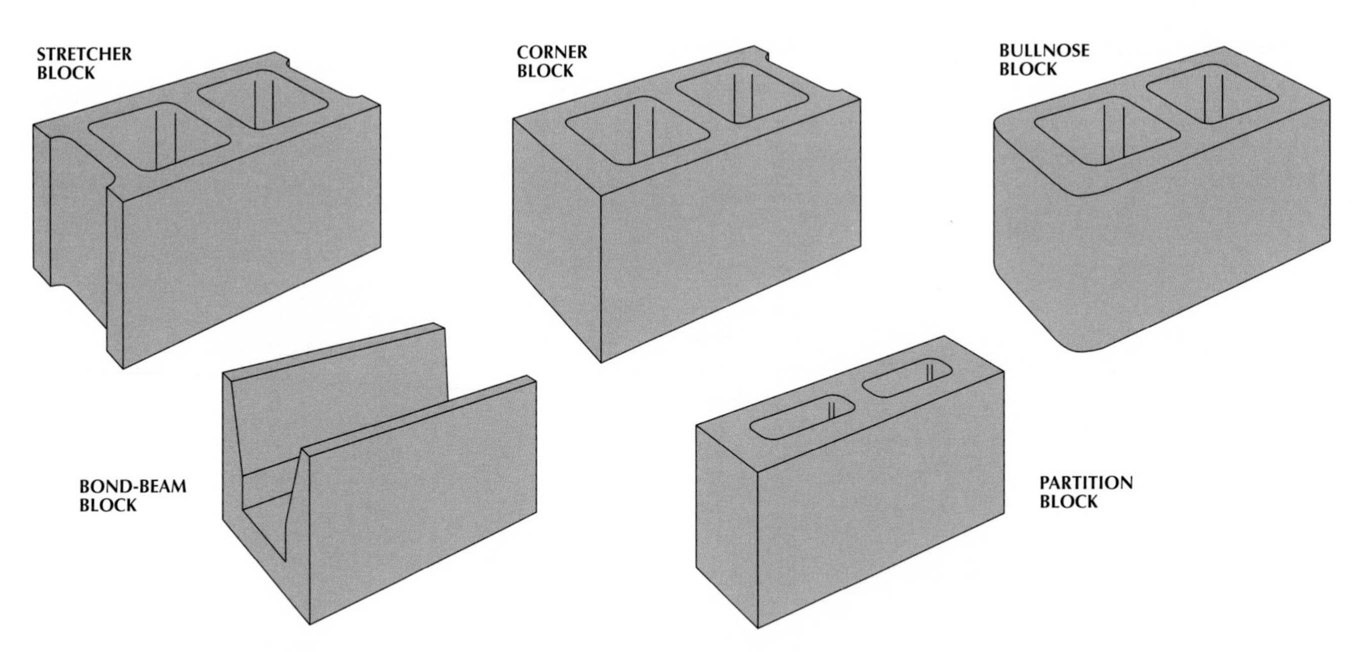


**Standard blocks.**
Most concrete-block walls are constructed with the five block types shown above, all of them variants of the standard hollow block. Stretcher blocks, with or without flanges at each end, are used for long runs of wall where both ends are mortared to adjacent blocks. (In areas prone to earthquakes, the type without flanges allows vertical reinforcement to run through the block cores more easily.) Most blocks have wider webs on one side to better accommodate mortar. Corner blocks have one flat end for smooth surfaces at corners and wall ends. Bullnose blocks, for similar applications, have one or two rounded corners. Channel-shaped bond-beam blocks are used to produce a continuous cavity along the top of a wall, which is filled with horizontal steel reinforcing rods and grout. The bond-beam blocks sold in seismic areas have cores to accommodate vertical rods as well. Partition blocks are half the width of a standard block and are used for interior partition walls, cavity walls, and composite walls.

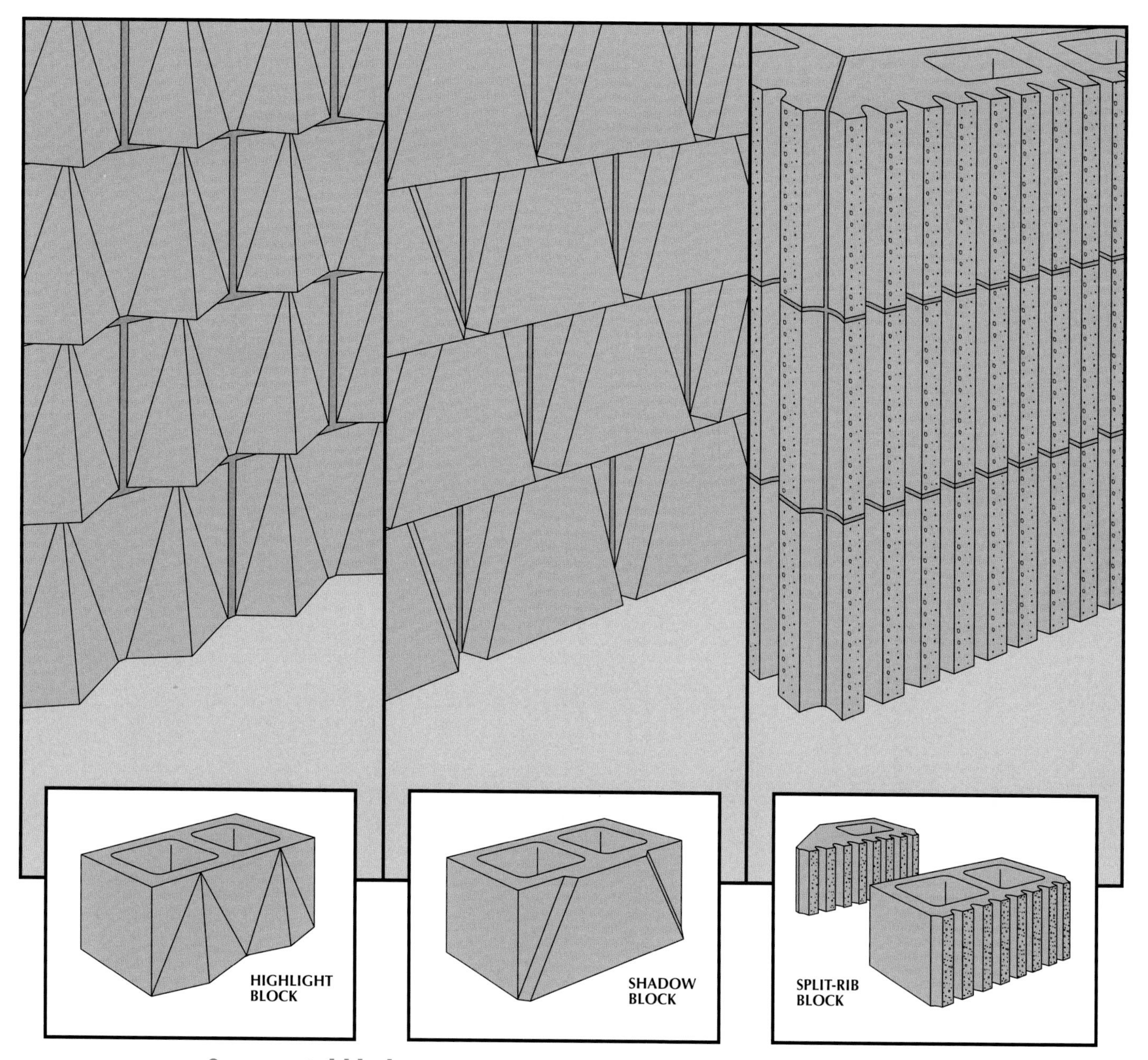


**Ornamental blocks.**
Faceted blocks are designed to create an interplay of light and shadow in a wall. They are available in a wide range of patterns, textures, and shapes, and are manufactured in the standard nominal dimensions of 8 inches by 8 inches by 16 inches. Blocks of different dimensions are available, but usually need to be specially ordered.

Both highlight and shadow blocks can be used to produce a dramatically sculpted wall surface. Split-rib blocks yield an array of narrow vertical columns, which can be continued around a corner with the use of special matching blocks with a mitered edge. The variety of aggregates used in the concrete recipes for split-rib blocks gives a selection of textures.

# Selecting and Shaping Stone

Only a few hand tools are necessary for the relatively simple task of fashioning raw stone into building blocks.

**Obtaining Stone:** Most stone used for construction falls into a half-dozen major types *(chart, opposite)*, and comes in two basic forms—as fieldstone or as quarried stone. Fieldstone tends to be rounded and weathered. You can buy it from a building-materials supplier or gather it yourself *(pages 17-18)*. Quarried stone has a fresh-cut appearance and is generally easier to split and shape.

Building-materials suppliers often ship in stone from distant sources, and must necessarily raise prices to cover transport. Local stone, therefore, is generally cheaper. Most quarries will deliver it directly to the building site. You can cut costs by hauling it yourself in a rented truck or trailer.

**Shaping Techniques:** Before starting a cut, study the stone to determine its grain, evidenced by layers or striations, and plan to capitalize on this direction of natural splitting. If the stone has no grain, its cutting characteristics will be affected only by the density of the stone. In general, a dense stone such as granite requires many more chipping blows to achieve a shape than does the less dense slate or sandstone.

Small stones can be hand-held during the shaping process, but place heavier pieces either on a low, sturdy workbench or on the ground. Cushion the bench or ground with padding, sand, or sawdust to absorb the force of the blows. Lack of cushioning may cause the stone to break at the point where it touches the work surface.

**Tools:** The mason's basic shaping tool is a special hammer with a head that is blunt at one end and wedge-shaped at the other. It is used for breaking and splitting large stones or for chipping edges. The best weight for general use is 3 pounds. A maul, blunt at both ends, is used to strike chisels in dressing the face of a stone and to drive the wedge-shaped end of the stone hammer into stubborn splits. Three pounds is also a good all-purpose weight for a maul. Useful for quick, fine shaping of soft stones is a bricklayer's hammer with a chipping blade. The basic set of chisels shown below will suit most jobs.

**TOOLS**

- Maul
- Stone chisel
- Stone hammer
- Try square
- Pitching tool
- Pointing tool
- Bricklayer's hammer

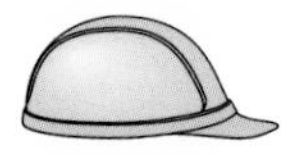

**SAFETY TIPS**

*Whenever cutting or dressing stone, wear leather work gloves, a long-sleeved shirt, and goggles. Hard-toed shoes prevent injury from falling stones.*

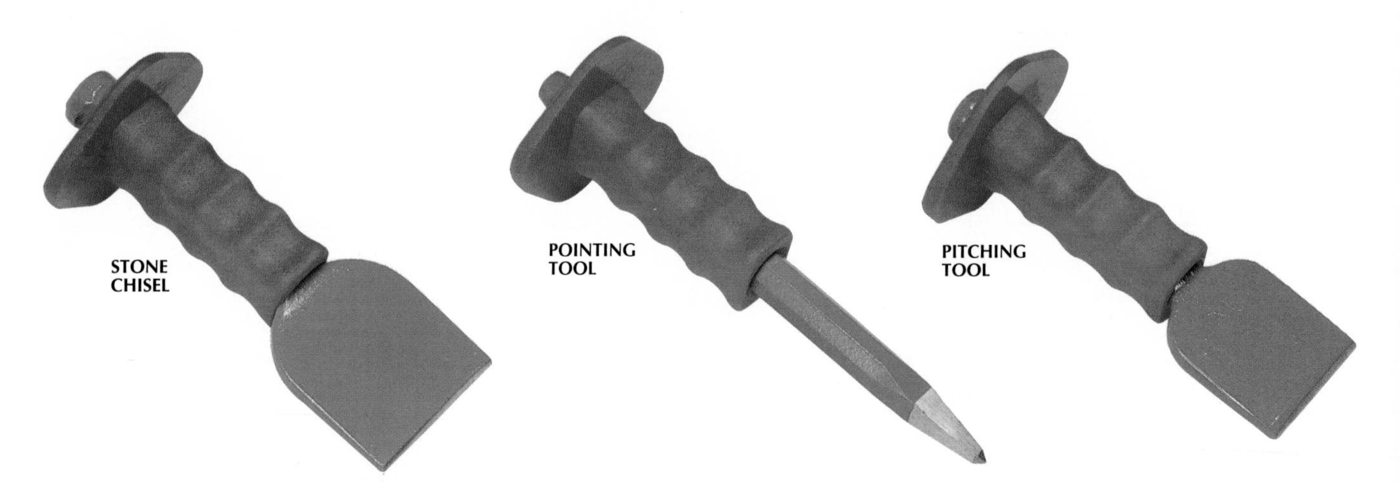


**Stone chisels.**
A basic set of chisels for stone work includes a sharp, wide-blade stone chisel for splitting, cutting, and notching; a sharp-tipped pointing tool for fine dressing; and several widths of heavy, blunt-tipped pitching tools for shearing off small protrusions.

## A GUIDE TO COMMON BUILDING STONES

| Type | Durability | Workability | Water Resistance | Weight | Color | Texture | Uses |
|---|---|---|---|---|---|---|---|
| **Granite** | Good | Difficult | Good | Heavy | Various grays | Fine to coarse | Building |
| **Basalt** | Excellent | Difficult | Excellent | Heavy | Black | Fine | Paving |
| **Limestone** | Fair | Medium to difficult | Poor | Heavy | Various | Fine to coarse | Building, veneering |
| **Slate** | Good | Easy | Excellent | Medium | Purple, gray, green | Fine | Veneering, paving, roofing |
| **Shale** | Poor | Easy | Poor | Medium | Various | Fine | Veneering |
| **Sandstone** | Fair | Easy to medium | Fair | Light to medium | Various | Fine to coarse | Building, veneering |

**Selecting stone.**
This chart compares the properties of several common types of construction stone. Stones in each type can vary greatly, however, depending on the region where they are obtained. Durability and workability are the main factors to consider in choosing a type for a project. A coarse-grained sandstone will crumble, for instance, if it is used for a load-bearing pier or in a wet, exposed location. Dense granite and limestone are suitable for almost any application, but can be difficult to split. If you will be cutting a great deal of stone into thin slabs for paving or veneer, look for a type with parallel grain—natural break lines that will make splitting easier. If your area has severe freezing and thawing, choose a type that is highly water resistant.

# SPLITTING STONE

**1. Chiseling with the grain.**
◆ Position a wide-blade stone chisel in line with the stone's natural grain. Lightly tapping the chisel with a maul, work your way across the stone to cut a shallow groove *(above)*.
◆ Return to the starting point and strike the chisel with more force, once again traversing the line. Repeat until the stone splits. If the stone cracks but does not split, you will have to use a stone hammer *(Step 2)*.

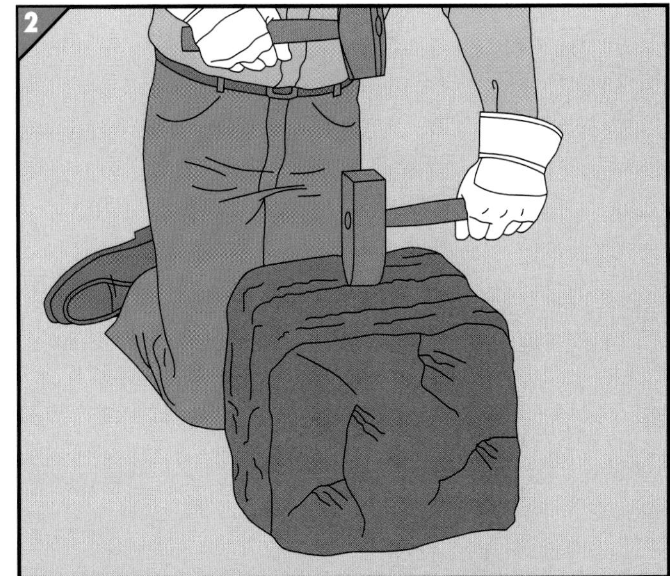


**2. Splitting stone with a stone hammer.**
◆ Align the wedge-shaped end of a stone hammer along the crack, then strike the flat face of the hammer firmly with a maul *(above)*.
◆ Move the stone hammer little by little along the crack, striking it repeatedly until the stone splits apart.

# SQUARING THE FACES OF A CORNER STONE

### 1. Rough-chipping a flat face.

◆ Scribe the cutting line on the stone with the corner of a stone chisel.
◆ Turn the stone so the waste edge is facing you.
◆ Chip off small pieces from the waste edge by striking the stone with glancing blows from the blunt end of the stone hammer; hold the hammer at a slight angle so that only the edge of the hammer face strikes the stone *(right)*.
◆ Continue chipping until you have reduced most of the surface unevenness and are within $\frac{1}{2}$ inch of the cutting line. Check the work occasionally with a try square to be sure the new face is perpendicular to the adjoining faces of the stone.

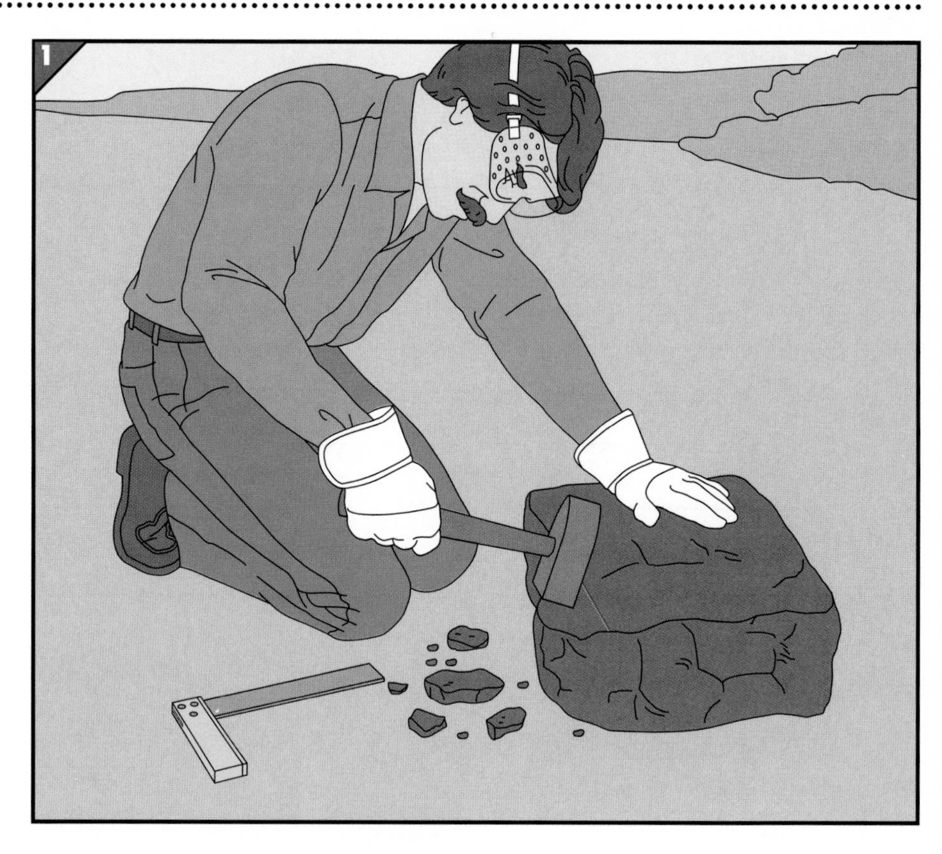

### 2. Smoothing the cut edge.

◆ Turn the stone on end.
◆ Holding a 2-inch pitching tool at about a 30-degree angle to the face of the stone, strike it with a maul. Continue chiseling in this manner, cutting a little at a time and working from the ends toward the center; turn the stone as required.
◆ When only small protrusions remain, remove them with a pointing tool, following the same chiseling technique *(left)*.

## BREAKING ACROSS THE GRAIN

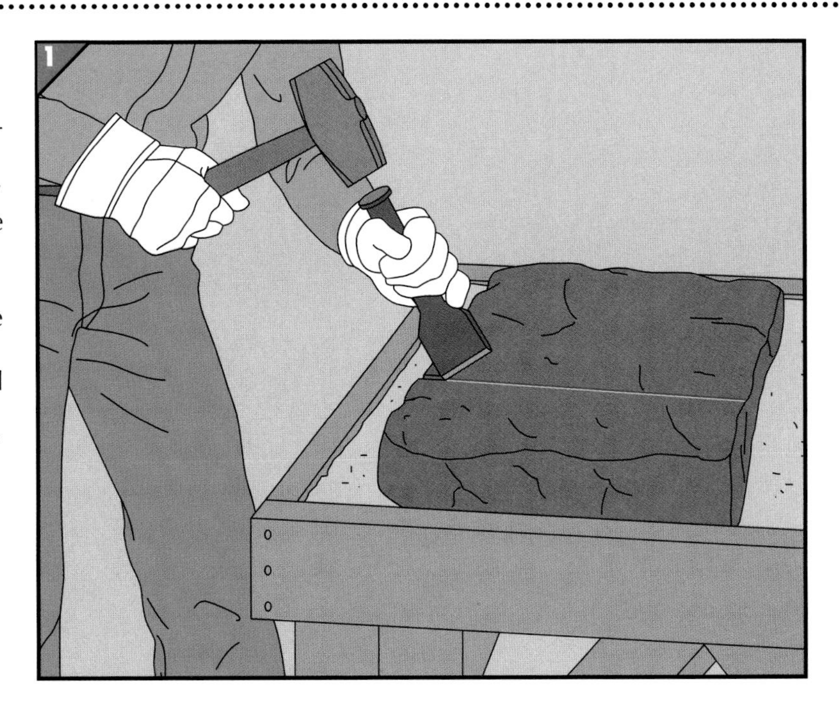


**1. Cutting a groove.**

◆ Scribe a cutting line on the stone with the corner of a wide-blade stone chisel.

◆ Holding the chisel so that only a corner of its blade contacts the stone, tap it lightly with a maul, following the scribed line to cut a shallow groove across the face of the stone *(right)*. Repeat this cut, each time tapping the chisel more firmly, until the depth of the groove is about one fifth of the stone's thickness.

◆ Turn the stone over and groove the other face in the same way, lining up the grooves as accurately as possible, then groove each edge.

**2. Snapping off the waste.**

◆ Align the groove with the edge of the workbench or a sturdy board.

◆ Press down on the edge of the stone with your free hand and strike the overhanging waste edge with a stone hammer, bringing the full face of the hammer down sharply against the surface *(above)* until the waste piece breaks off.

**3. Trimming the cut edge.**

◆ Stand the stone upright, with the cut edge facing you.

◆ Remove any large protrusions with a pitching tool and a maul *(opposite, Step 2)*.

◆ Working toward the center of the stone, chip away small uneven parts with the chisel end of a bricklayer's hammer by striking across the base of each protrusion *(above)*.

# CHIPPING STONE INTO A CURVE

**1. Scoring the curve.**

◆ Make a template of heavy cardboard cut to the desired curve and place it on the face of the stone.

◆ Scribe a cutting line on the stone with the corner of a wide-blade stone chisel, following the cardboard guide *(right)*.

◆ Mark the opposite face of the stone in the same manner, aligning the template with the first cutting line at the edges of the stone.

◆ Cut grooves along both cutting lines with a wide-blade stone chisel and a maul *(page 15, Step 1)*.

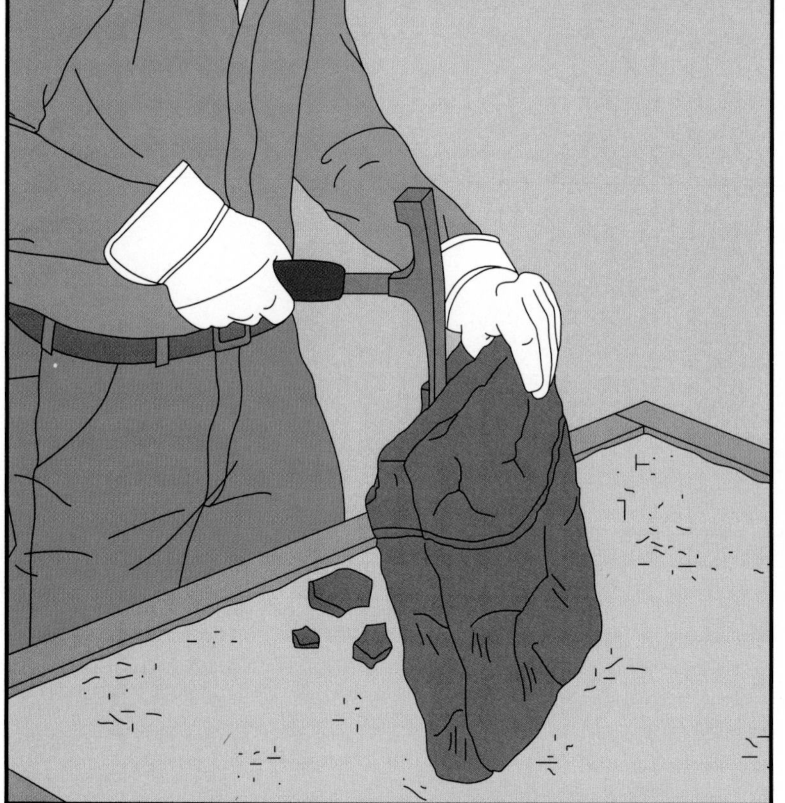

**2. Removing the waste.**

◆ Working along the edge of the waste, undercut the stone by chipping out large flakes—about half the thickness of the stone—with the chisel end of a bricklayer's hammer *(left)*.

◆ When the entire edge has been undercut in this manner, tap the overhanging edge with a stone hammer to snap it off.

◆ Repeat this procedure on the new edge of the waste, continuing to remove the stone in flakes until you reach the groove defining the curve.

◆ Smooth the curved edge with a pitching tool and a pointing tool *(page 14, Step 2)*.

# Salvaging Stone

You may be able to gather stone at no cost if you have access to rocky land or a stony creek. Farmers are often happy to be rid of stones plowed from their fields, and the sides of newly graded roads frequently abound in stones unearthed by bulldozers. Abandoned mines and quarries, shown on geological survey maps, may be surrounded by usable stones. Ruined stone chimneys, foundations, and walls also often contain stones of proven utility. Always check with the property owner before collecting such free stone, as it may have been earmarked for another use, or it could have historical value.

**Gathering Stones:** The best tool for stone gathering is a 30-pound, 5-foot-long digging bar, available at building-supply houses. It functions primarily as a lever for prying stones from a wall or from the ground *(below)*, but can also be used to split large stones.

**Transporting Heavy Loads:** Because of the sheer weight of stone—one cubic foot of granite may weigh 175 pounds—transporting it calls for some special procedures. If you must haul the stone over a long distance, rent a truck or trailer designed for heavy loads, and take care not to overload the vehicle—doing so may make it difficult to steer. Distribute the stones evenly, and pile partial loads toward the front of the bed.

In preparing to load the stones, drive as close to them as possible. Use a sturdy wheelbarrow to move them to the vehicle, or build an old-fashioned stone boat *(page 18)*. You will need a helper to lift large stones.

**TOOLS**

| | |
|---|---|
| Pick | 2 x 6 planks for levers |
| Shovel | Wheelbarrow |
| Digging bar | |

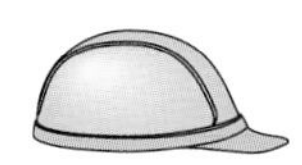

**SAFETY TIPS**

*A fresh cache of fieldstones may hide insects, rodents, or snakes. Wear heavy leather gloves to protect your hands from bites and stings, and from the jagged edges of stones. Put on hard-toed shoes to protect your feet.*

## EXCAVATING A LARGE FIELDSTONE

**1. Digging out a stone.**

◆ With a pick and a shovel, dig a trench around the stone deep enough to slip a 5-foot digging bar under it, leaving the earth piled beside the trench.

◆ Place a sturdy flat rock on top of the pile of earth to act as a fulcrum for the bar. Slide the bar under the stone and push down on the outer end to free the stone, while a helper slips a rock beneath the stone to support it *(left)*.

◆ Repeat the levering process on the other side until the stone is completely free and rests on its rock supports.

**2. Removing the stone.**

◆ Place a wheelbarrow at the edge of the hole.

◆ Slide two long, 2-by-6 planks as far as possible under the stone.

◆ Keeping the planks close together and parallel, work the stone onto them by prying and pushing from the opposite side with the digging bar.

◆ Bracing the stone with the bar to prevent it from rolling back into the hole, have your helper push down on the outer ends of the planks to lift the stone to ground level *(above)*.

◆ From the side opposite the wheelbarrow, slide the bar under the end of the planks to support them while your helper maneuvers the stone along the planks and into the wheelbarrow.

◆ When the stone is partway into the wheelbarrow, pull the bar from under the planks and use it to lever the piece into the wheelbarrow.

TRICKS OF THE TRADE

### Building a Stone Boat

If the terrain is very rough, you can make a heavy sled, or stone boat, to help with the removal of large stones. Design the boat low enough to allow stones to be tumbled rather than lifted onto it. Build it of 2-inch lumber, with removable sides held by metal brackets, and with runners on the bottom, capped with steel strips if possible. Attach a chain or heavy rope at each end to pull the sled or lower it down steep slopes.

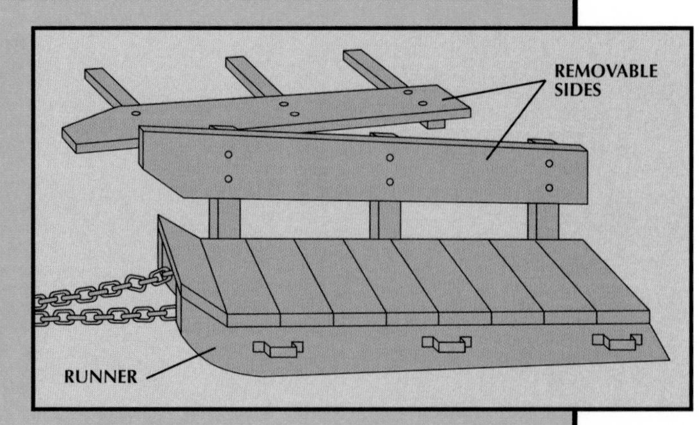

# Drilling and Cutting Masonry

Cutting through concrete and drilling holes in brick is hard work, but a number of special tools can make the job easier.

**Hand Tools:** The simplest tool for making a hole larger than 1 inch in diameter is a pointed chisel called a star drill that is struck with a maul *(below)*. Because the holes often end up with ragged edges, the star drill is best used where the final appearance is of little importance.

**Power Drills:** To make clean holes less than 1 inch in diameter, such as those needed to seat masonry fasteners *(pages 23-25)*, choose an electric hammer drill with a carbide-tipped bit *(page 20)*. This tool combines the twisting action of a regular electric drill with a hammer action. Small holes can also be made with a standard electric drill fitted with a carbide-tipped bit, but the work will be much slower.

To bore holes between 1 and 6 inches in solid concrete, the best tool is a rotary hammer *(page 21)*. It combines a drilling and a hammering action, but will drive a bit into masonry with much greater force than a hammer drill. Depending on the needs of the job, the rotary hammer can be switched from the combination drill-and-hammer action to a drill-only or a hammer-only mode. It can also be fitted with special chisel bits and used as a hand-held jackhammer to blast sizable openings in masonry walls. Fitted with other specialty bits, the rotary hammer can help with removing excess mortar and cutting around a brick so you can take it out of a wall.

**Masonry Saws:** For making a cut through masonry, the ideal tool is an abrasive saw *(page 22)*. Similar to a circular saw and fitted with a masonry blade, the abrasive saw applies greater power and uses a larger blade. The exact type of blade required depends on the kind of material being cut. The grit on the blade will wear down during use, so the blade requires periodic replacement.

**CAUTION** *Always unplug electric tools before adjusting them or changing their bits or blades. When working outdoors, do not work under damp or wet conditions.*

**TOOLS**

- Maul
- Star drill
- Hammer drill and bits
- Ruler
- Electric drill and bits
- Rotary hammer
- Bull-point chisel bit
- Scaling chisel
- Slotting chisel
- Abrasive saw
- Masonry blade
- Cold chisel

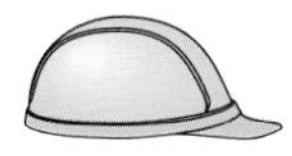

**SAFETY TIPS**

*Wear sturdy work gloves and goggles when drilling and cutting masonry. When working with power tools, add a dust mask.*

## ROUGH-CUT HOLES BY HAND

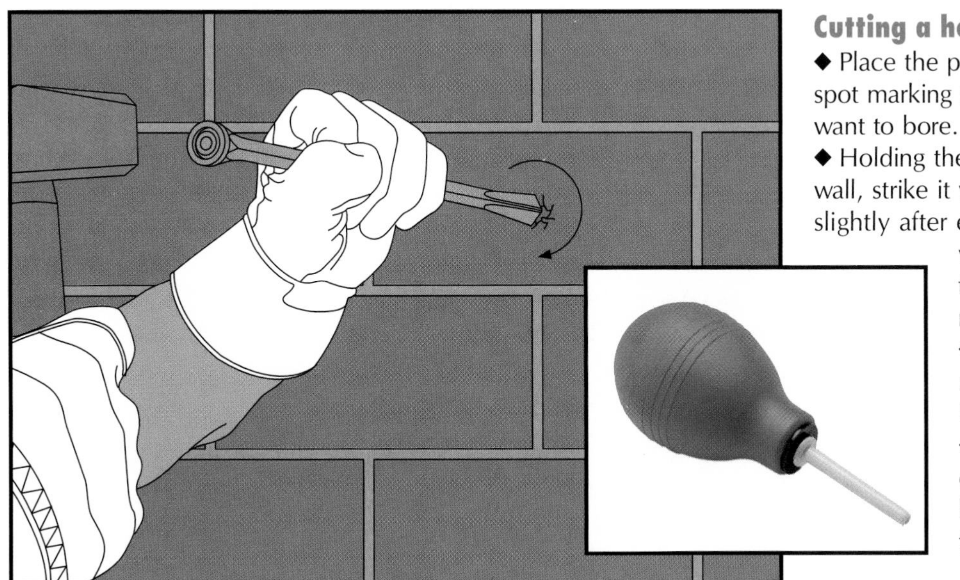

**Cutting a hole with a star drill.**

◆ Place the point of the star drill on the spot marking the center of the hole you want to bore.

◆ Holding the drill perpendicular to the wall, strike it with a maul. Rotate the drill slightly after each hammer blow to prevent the four cutting blades from getting stuck in the masonry *(left)*.

◆ Continue hammering and rotating the drill until the hole is the desired depth.

◆ After the hole is finished, clear it of dust and debris; a blowout bulb will simplify the job *(photograph)*.

# A HAMMER DRILL FOR SMALL HOLES

**1. Preparing the tool.**

◆ With the cord unplugged, fit a bit of the required diameter into the drill; choose a bit at least 1 inch longer than the desired depth of the hole.
◆ Lay the drill on a flat surface and loosen the wing nut that secures the depth gauge.
◆ Place a ruler under the bit, aligning the end of the ruler with the tip of the bit.
◆ Slide the depth gauge forward or backward until the end of the gauge lines up with the mark on the ruler corresponding to the desired depth of the hole *(right)*.
◆ Tighten the wing nut to secure the gauge.

If the hammer drill has no depth gauge, wrap the bit with tape at the desired depth of the hole.

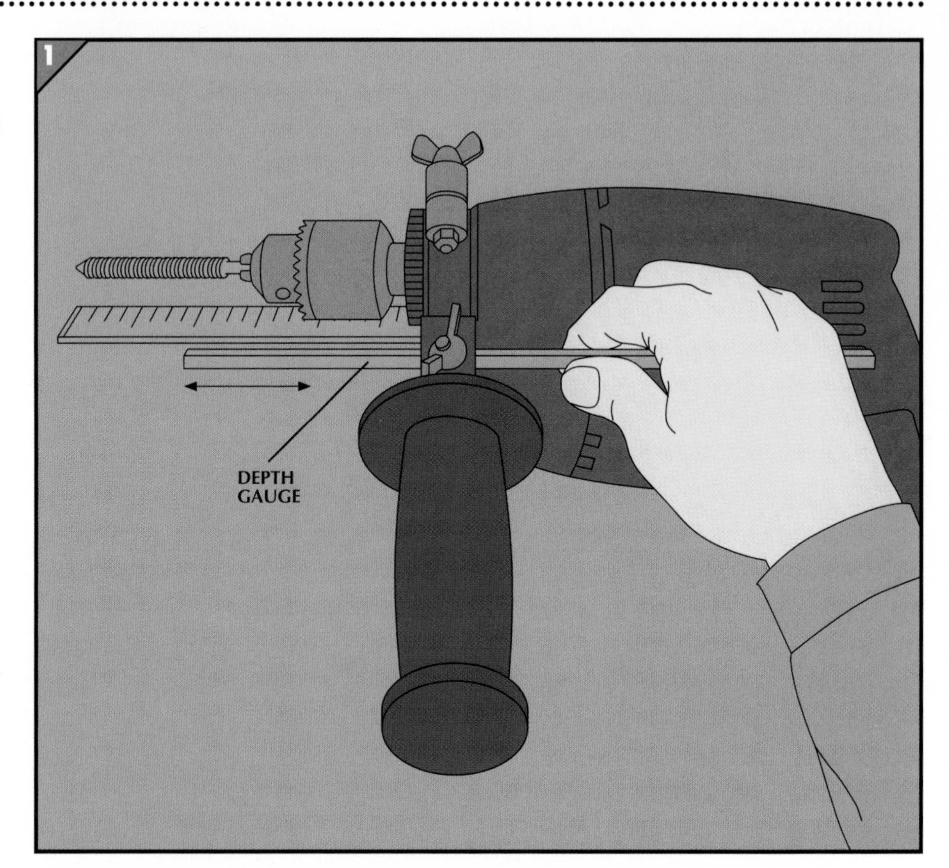


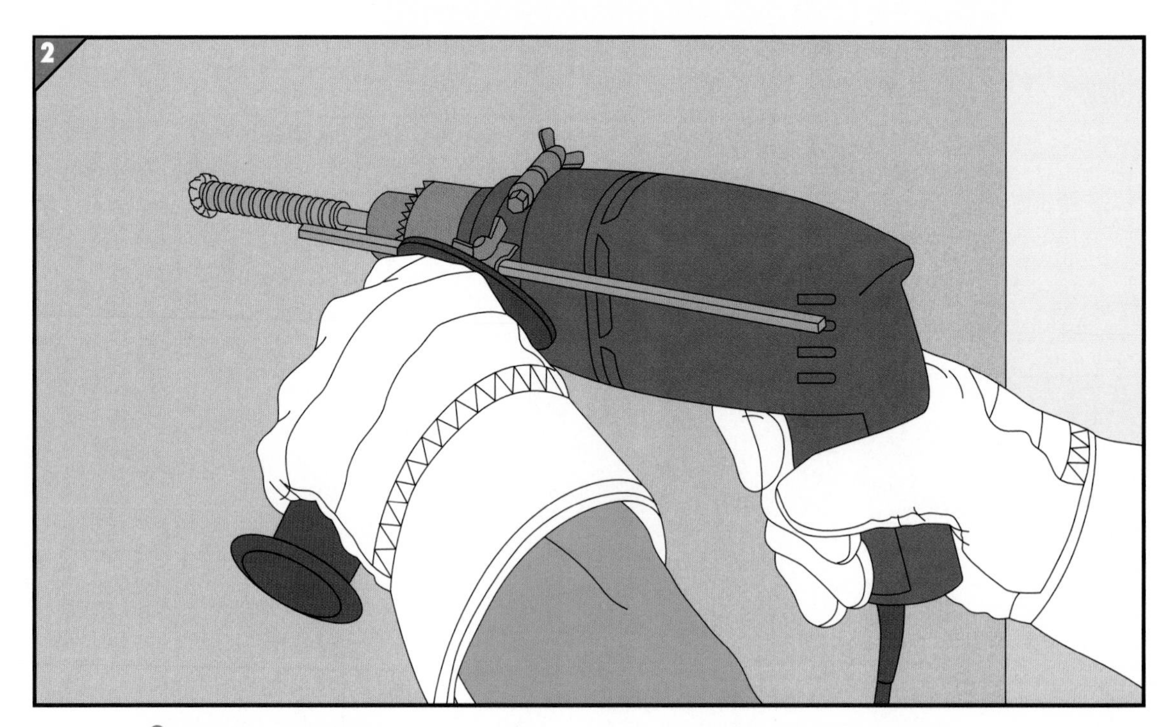


**2. Boring the hole.**

◆ Mark the center of the planned hole.
◆ Touch the point of the bit to the mark, then turn on the hammer drill and press it forward, applying just enough pressure to keep the bit in contact with the masonry *(above)*.
◆ Continue drilling until the depth gauge or tape touches the surface of the wall.

**Chiseling through concrete.**

◆ Mark the outline of the proposed opening, and notch a starter hole in its center with a star drill *(page 19)*.
◆ Set the rotary hammer for HAMMER ONLY and attach a bull-point chisel bit, following the instructions appropriate to the model.
◆ Grip the tool firmly in both hands and touch the bit to the starter hole.
◆ Turn on the hammer and drill straight into the hole *(right)*.
◆ After you have penetrated about 4 inches, turn off the machine, withdraw the bit, and reposition it at the edge of the hole. Turn on the hammer again and widen the cut.
◆ Continue in this manner until you reach the outline, resting the hammer occasionally to prevent it from overheating.
◆ Rework the cut area, deepening it until you pierce the concrete.

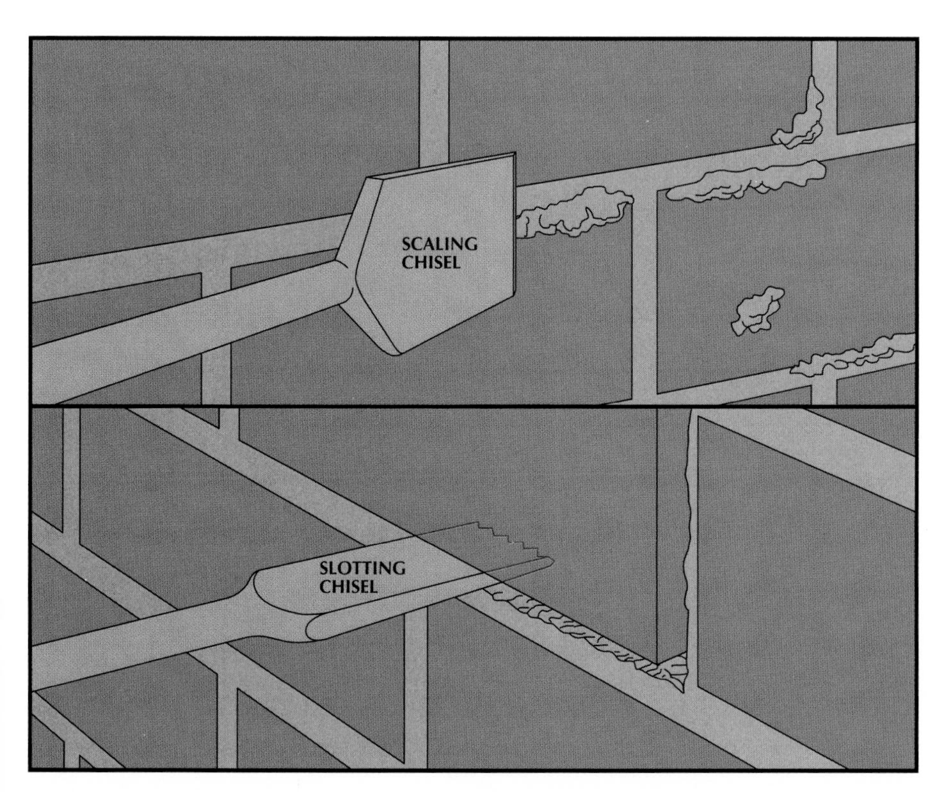


**Removing unwanted mortar.**

◆ To remove spattered mortar from brick, set the rotary hammer to HAMMER ONLY and fit it with a scaling chisel.
◆ Hold the chisel almost parallel to the wall and guide it along the surface so that the bit lifts off the debris without gouging the faces of the bricks *(left, top)*.

◆ To remove one or more bricks from a wall, set the rotary hammer to HAMMER ONLY and fit it with a slotting chisel.
◆ Insert the serrated blade of the chisel into the mortar joint and cut a narrow slot all the way around the brick, until the brick can be removed *(left, bottom)*.

# AN ABRASIVE SAW FOR BRICK AND BLOCK

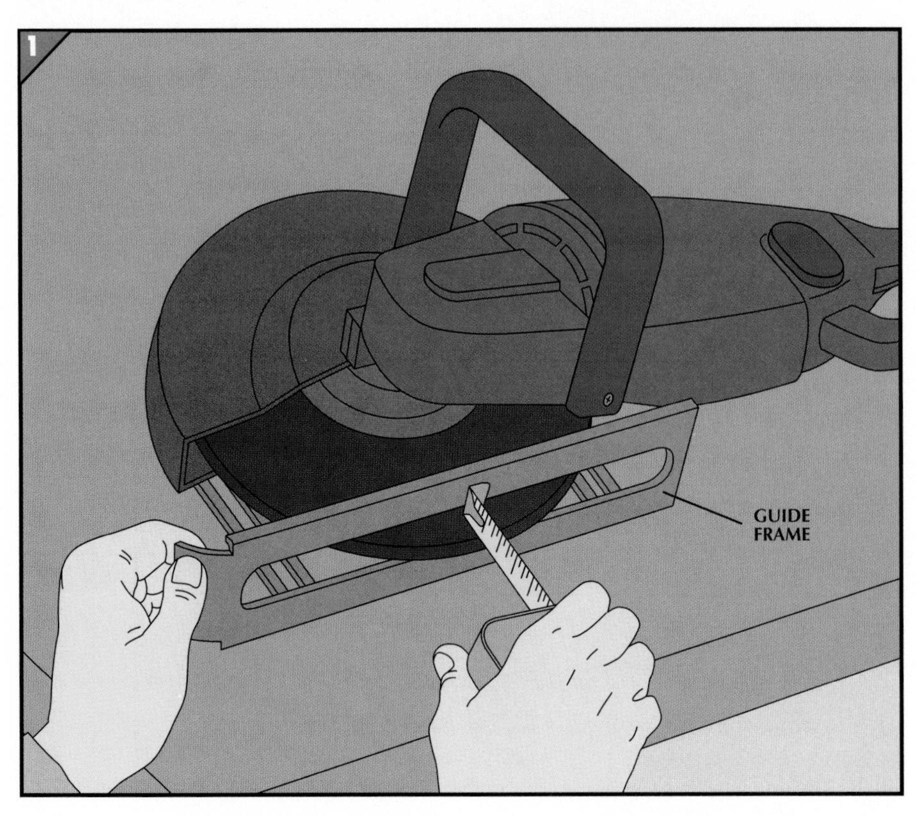


**1. Adjusting the depth of the cut.**

◆ With the saw unplugged, loosen the nuts holding the guide frame in place.

◆ Tilt the saw on its side and adjust the guide frame until the distance between the frame and the bottom of the blade equals $\frac{1}{2}$ inch *(left).*

◆ Tighten the nuts to secure the guide frame in this position.

CAUTION *Before using the saw, test the blade for invisible cracks by suspending it on a pencil and tapping it with a screwdriver handle. If it rings with a sharp bell tone, the blade is good. Do not apply sideways pressure when cutting with the saw—doing so may cause the blade to shatter.*

**2. Operating the saw.**

◆ With a pencil or a masonry crayon, outline the desired opening on the wall. Wherever possible, position the outline so that it follows mortar joints.

◆ Place the saw over the cutting line with the center of the blade a few inches in from one corner of the marked opening so it cuts close to, but not beyond, the corner.

◆ Turn on the saw and lower the blade onto the line until the guide frame touches the masonry.

◆ Move the saw forward very slowly *(right)* until the leading edge of the blade reaches the corner. Turn off the saw and lift it from the cut.

◆ Following the same procedure, make identical cuts on the other marked sides.

◆ Reset the guide frame to allow a cut $\frac{1}{2}$ inch deeper, and retrace the original cut.

◆ Saw until you have reached the desired depth, or until you have cut as deep as the saw will go—usually $4\frac{1}{2}$ inches.

◆ With a cold chisel and a maul, complete the cut at the corners, where the blade penetrated less deeply, then break up the concrete within the outline with the maul and remove it.

# Fasteners for Masonry

When choosing among the many masonry fasteners available, it is important to consider both the load on the fastener and the composition of the masonry itself.

For almost every fastening job there are several suitable fasteners *(page 25)*. The simplest of these are nails, pins, and screws, which are inserted directly into the masonry. Although these fasteners serve for most jobs, they provide only light holding power in soft masonry, and are difficult to drive into hard masonry such as granite or old, dense concrete. For these surfaces, and for mounting heavy loads on any surface, you will need a two-part fastening system.

**Nails, Pins, and Screws:** Made of hardened steel, masonry nails can be driven with a heavy hammer. A magnetic nail starter, or punch, steadies nailing in tight places *(below)*. Steel pins, with metal washers set behind their points, are a bit lighter than nails but are longer and can penetrate the surface more deeply. Pins are driven with a stud driver, a tool consisting of a protective barrel, which holds the pin, and a piston, which fits into the barrel and rams the pin into the masonry. When the steel pin is fully set, its metal washer will prevent mounted fixtures from working their way loose. For small jobs, a manual stud driver, used with a hammer, serves well enough *(page 24, top)*; but for driving many pins, a powder-actuated stud driver is more efficient *(box, page 24)*.

Masonry screws are often used where hammering could ruin the appearance of a surface. A pilot hole for the screw is drilled first; most manufacturers supply a drill bit with each box of screws to ensure that the hole will be exactly the right size.

**Two-Part Fastening Systems:** A screw, nail, or bolt fixed in an anchor or shield of plastic or metal provides a solid hold in masonry. The anchor is inserted into a hole drilled in the surface *(pages 19-21)*, and the fastener is then driven into the anchor, expanding it to wedge the assembly in place.

Another type of anchor, called a toggle, is used in hollow walls; it consists of a screw or a bolt fitted with a pair of retractable wings that open and grip the inside wall surface when the screw or bolt is tightened.

**TOOLS**

Nail starter
Ball-peen hammer
Stud driver
Maul

**MATERIALS**

Masonry nails
Steel pins

**SAFETY TIPS**

*Wear goggles when driving fasteners into masonry surfaces.*

## DRIVING NAILS AND PINS

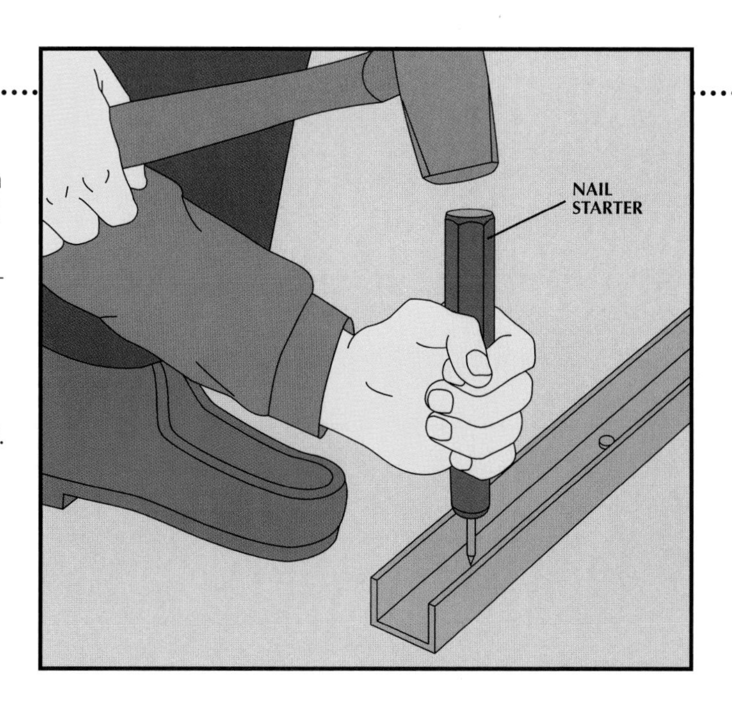


**Nailing in close quarters.**

◆ Center the head of a masonry nail on the circular, magnetized end of a nail starter, then position the point of the nail against the object being mounted—in this case, a metal channel being fastened to a concrete floor.

◆ Tap the end of the nail starter lightly with a maul or a heavy ball-peen hammer to break the masonry surface *(right)*.

◆ Drive the nail by striking it with heavier blows.

**Setting a steel pin.**

◆ Slip the head of a steel pin into the barrel of a stud driver *(above, left)*, pushing the pin into the barrel until the metal washer behind its point touches the end of the barrel.

◆ With the pin in place in the stud driver, press the point of the pin against the object being mounted on the masonry—here, a furring strip.

◆ Grasp the handgrip of the stud driver firmly and strike the driver piston with a maul, forcing the pin through the washer and into the masonry *(above, right)*.

## A POWDER-ACTUATED STUD DRIVER

A quick and easy way to drive steel pins into a masonry surface is with a powder-actuated stud driver. This tool—which is literally a kind of gun—uses the explosive force of a .22-caliber blank to drive the piston against the pin. It must therefore be handled with great care; if you are renting the tool, be sure to get the manufacturer's instructions for the exact model you are renting, as its operation may vary from one model to another. Always wear goggles and earplugs when you are working with the stud driver.

CAUTION *Carefully follow all safety precautions that come with the tool, and read the instructions to determine whether the tool is appropriate for the type of surface you will be fastening into—some materials can shatter or cause the pin to rebound. Keep the tool out of the reach of children.*

## SUITING THE FASTENER TO THE SURFACE AND THE LOAD

| Type of Fastener | Installation Method | Application | | | | | | | | |
|---|---|---|---|---|---|---|---|---|---|---|
| | | Adobe | Block, core | Block, solid part | Brick | Concrete, | Concrete, dense | Mortar joints | Stone, hard | Stone, soft |
| Masonry nail | Choose a nail that will penetrate masonry $\frac{1}{2}$ to $\frac{3}{4}$ inch. Tap lightly to start; then pound home. | | | L | L | H | H | L | | L |
| Steel pin | Drive with a stud driver *(opposite)*. | | | L | L | H | H | L | | L |
| Masonry screw | Drill hole with bit provided. Drive screw through object being mounted and into hole. | L | | L | L | L | H | L | | |
| Plastic nail anchor | Drill hole for anchor through object and into wall. Insert anchor and drive nail. | L | L | L | L | L | L | L | L | L |
| Plastic screw anchor | Drill hole the diameter of anchor. Tap anchor into hole until flush with surface; position object and drive screw. | L | L | L | L | L | L | L | L | L |
| Metal nail anchor | Drill hole the diameter of anchor through object and into wall. Position object, insert anchor, tap lightly to seat, then drive nail. | | L | H | H | H | H | H | H | L |
| Lag shield | Drill hole, insert shield, position object, and drive lag screw. In mortar joints, set shield so that it expands against the masonry units. | L | | H | H | H | H | H | H | L |
| Sleeve anchor | Drill hole the diameter of anchor through object and into masonry. Insert screw or bolt into anchor, tap into hole, and tighten. | H | | H | H | H | H | H | H | H |
| Hammer-set anchor | Drill hole the depth and diameter of anchor. Insert anchor until flush with surface, then tap with setting tool to seat. Insert bolt through object and tighten. | | | H | H | H | H | H | H | H |
| Plastic toggle | Drill hole to fit folded toggle. Flatten anchor and push into hole. Position object, insert screw, and tighten. | | L | | | | | | | |
| Metal toggle | Drill hole to fit folded toggle. Push bolt through object; add toggle. Insert toggle into hole, pull to hold against inside of wall, and tighten bolt. | | H | | | | | | | |

**Choosing a masonry fastener.**
The chart above shows some of the masonry fasteners available for home use, and explains how each kind is installed. The surfaces for which each kind of fastener is appropriate are indicated on the far right. "H" means that the fastener or anchor is sturdy enough to withstand the pressures of virtually any kind of household structure—including bookshelves and stair handrails—provided the surrounding masonry is sound. "L" indicates that the fastener or anchor should be used for light loads only.

# Breaking Up Masonry

Most concrete slabs and masonry walls can be dismantled with perseverance and the proper tools; however, removing a wall that abuts a bearing wall or a slab that adjoins a foundation may have structural implications—consult an architect or an engineer first. Leave dismantling load-bearing walls, reinforced concrete block walls, and cast concrete walls to a professional. Before undertaking any demolition job, call the local building department to determine whether your project is covered by code.

**Concrete Slabs:** For a relatively small job, breaking up concrete can be done with a sledgehammer. Large stretches are more quickly demolished with an electric jackhammer *(below)*. With either tool, begin at the edge of the slab and work inward. How easily the concrete shatters will depend on its thickness, age, and quality.

Demolition cement is another alternative for breaking up concrete. When mixed with water and poured into pre-drilled holes in the surface, this product expands to fracture the concrete within a few hours. Always follow the manufacturer's instructions for its use.

**Brick, Block, and Stone:** To break up a brick wall with sound mortar joints, chip away at the mortar with a cold chisel and a maul *(opposite)*, working outward from a spot between pilasters, if the wall has them, and from the top down. Walls with old, cracked, or crumbling joints can be more easily toppled with a wrecking bar *(page 28)*. Block and stone walls can be dismantled using the same methods.

If you wish to reuse the masonry units you've removed, you will need to clean them *(page 28, Step 3)*. Pieces to be remortared in a wall must be fairly flat but need not be entirely clean. Bricks to be laid flat in a patio must be free of mortar chunks; mortar film will wear off with time.

**TOOLS**

- Jackhammer
- Bull-point bit
- Cold chisel
- Maul
- Bricklayer's hammer
- Wire brush
- Wrecking bar
- Sledgehammer

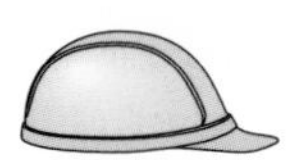

**SAFETY TIPS**

*For breaking up masonry, put on goggles and hard-toed shoes. Add hearing protection and a dust mask when using a jackhammer. To work with demolition cement, wear rubber gloves. Use a stepladder or scaffolding* (pages 29-31) *to keep your head and shoulders safely above falling debris.*

## DEMOLISHING A CONCRETE SLAB

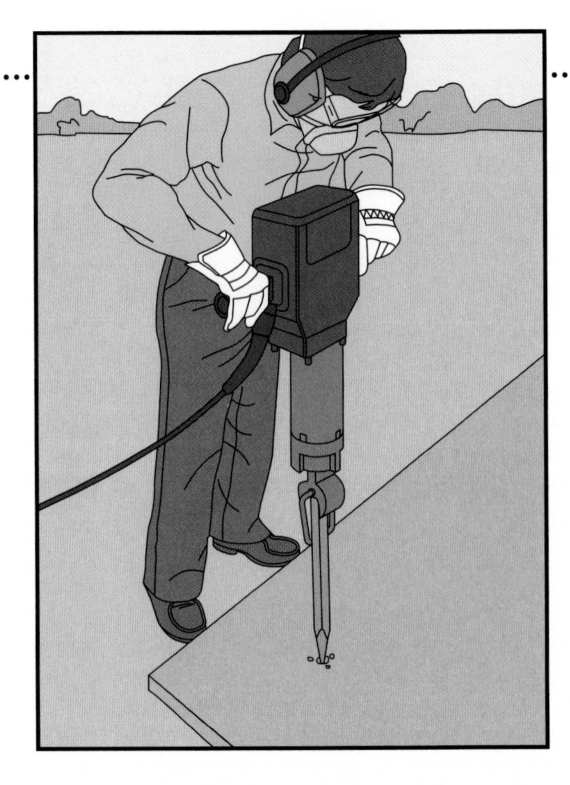

### Using a jackhammer.

◆ Install a bull-point bit in the jackhammer.

◆ Begin at a corner of the slab, positioning the bit against the concrete 1 foot in from both edges *(right)*. Press the trigger on the hammer handle and hold the hammer steady until the bit penetrates the slab to a depth of about 1 inch.

◆ Stop the hammer and drill another hole 1 foot in from the edge and 1 foot from the first hole. Chop several more holes in this way.

◆ Go over the row of holes you have just chopped several times with the jackhammer, drilling 1 inch deeper each time until you have penetrated the concrete completely.

◆ If the concrete does not fracture, reduce the spacing between holes to 6 inches. Continue reducing the spacing until the entire area you have drilled fractures.

◆ To break up the rest of the slab, drill holes at the spacing you've determined, working inward from the edge of the slab and never drilling more than 1 inch deep at a time.

**CAUTION** *Because of the noise and vibration caused by the electric jackhammer, never work with the tool for more than half an hour at a time.*

# DISMANTLING A WALL ONE COURSE AT A TIME

**1. Loosening mortar.**

◆ Starting in the middle of the top course of the wall, use a maul to pound a cold chisel deep into the vertical mortar joints at one end of a brick, driving the chisel into the joint repeatedly until all the mortar has been loosened *(left)*.

◆ Knock the loose mortar out of the joint with the chisel point.

◆ Repeat the procedure to chip out the mortar from the vertical joint at the other end of the brick.

◆ Remove the mortar in the horizontal joint below the brick in the same way.

**2. Removing a loosened brick.**

◆ Wedge the tip of the chisel into the loosened horizontal joint, angling the point downward.

◆ Pressing down on the chisel, pry up the brick. If the brick does not pop out, break it by striking it with the maul and pry it out in pieces, along with the surrounding mortar.

◆ Proceed to remove adjacent bricks in this way until the entire course is cleared *(right)*, then follow the same procedure to dismantle the other courses.

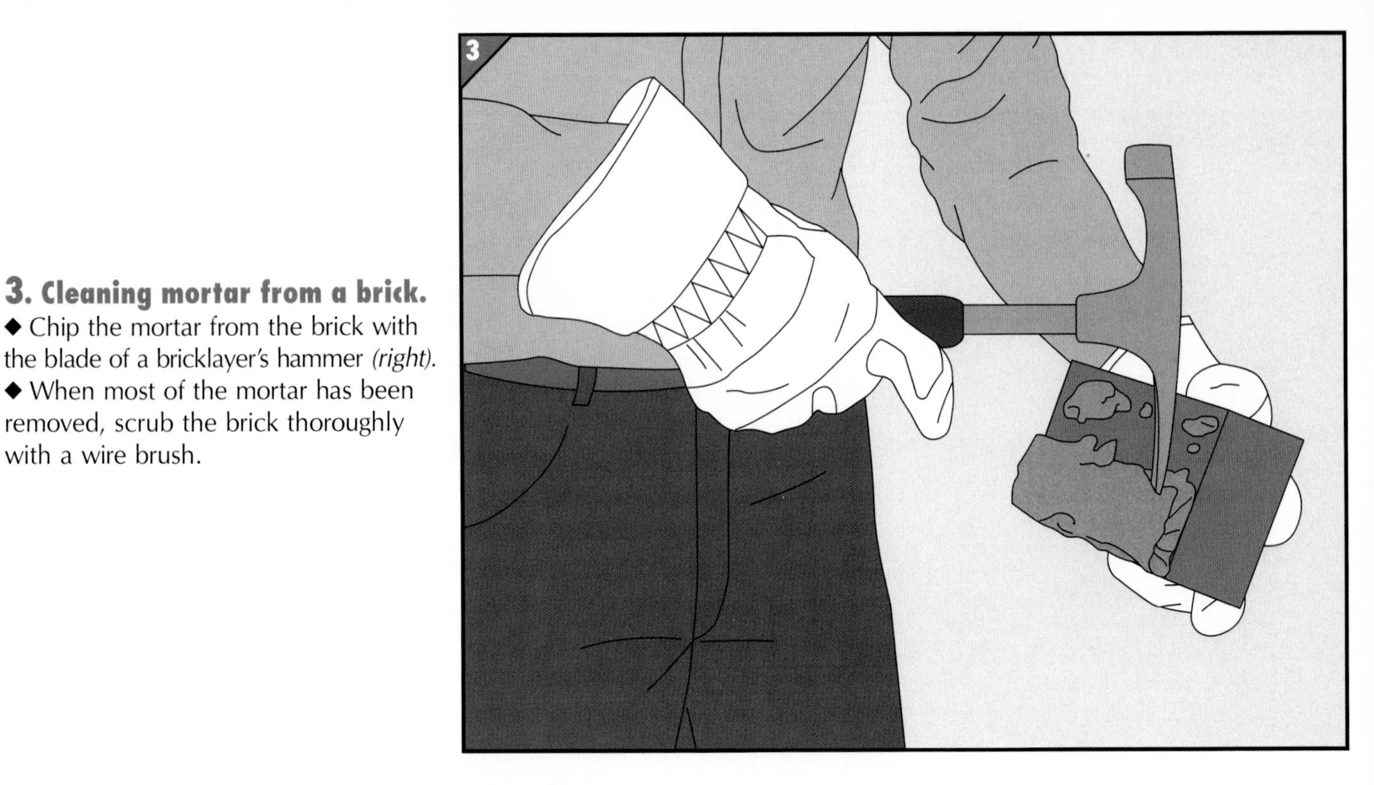


**3. Cleaning mortar from a brick.**

◆ Chip the mortar from the brick with the blade of a bricklayer's hammer *(right)*.

◆ When most of the mortar has been removed, scrub the brick thoroughly with a wire brush.

## TOPPLING AN ENTIRE SECTION

**Dismantling the wall.**

◆ Force the angled end of a wrecking bar into a cracked or crumbling mortar joint. If you are knocking down a double-thick wall that contains header courses interspersed with the stretcher courses—where bricks are laid crosswise over courses of bricks laid lengthwise—select a mortar joint under a course of header bricks.

◆ Push down on the curved end of the wrecking bar to loosen and dislodge the section of wall above *(left)*. If the wrecking bar does not move the bricks, remove it and strike the wall with a sledgehammer to loosen the mortar joints; direct the blows at the wall just above the point where you inserted the wrecking bar.

◆ Alternate prying with the wrecking bar and striking with the sledgehammer until the wall topples.

# Erecting Scaffolding

Scaffolding made of tubular steel sections that couple and lock together is safe and quick to put up. It can be rented with all the necessary hardware, and many suppliers will deliver the equipment to the job site and pick it up when the work is done. Be sure to get instructions from the rental agency about locking the frame pieces together correctly.

For the structure to be stable, it must be adjusted to the terrain *(page 30)*. Also, very high scaffolding—usually four or more frames tall—must be anchored to the wall to prevent tipping. To determine the anchor height required, measure the length and width of the scaffold, then multiply the smaller of these dimensions by 4. This figure equals the height at which the scaffold must be secured to the wall *(page 31)*.

TOOLS

Mason's level
Electric drill
Masonry bit
Wrench

MATERIALS

Scaffolding system
2 x 10s
Masonry anchors
Lag screws ($\frac{1}{2}$")
Tie bars
Tie brackets
Tie clamps

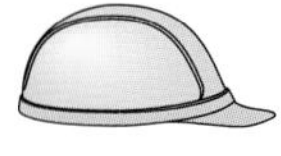

SAFETY TIPS

*Put on a hard hat when working on or around scaffolding.*

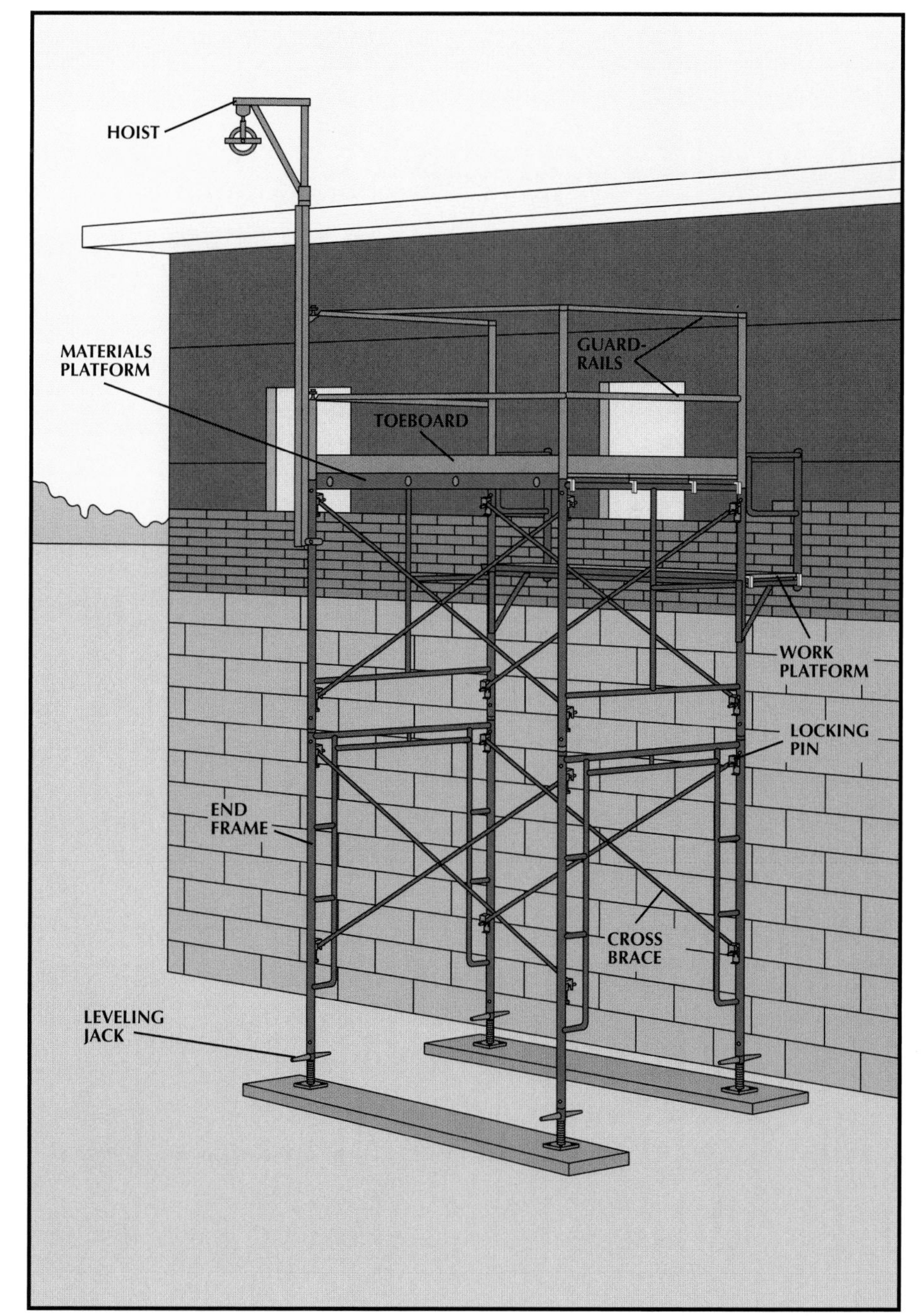


**Anatomy of a masonry scaffold.** Scaffolding consists of stackable, tubular-steel end frames held in place by cross braces and locking pins. End frames have footholds for climbing; adjustable jacks in the lowest end frames level the rig.

A second level supports a platform for materials. The optional work platform that is clamped to the side of the scaffolding frame brings the mason closer to the rising wall. Planks for both the materials platform and work platform consist of aluminum frames covered with plywood that hook over the end frames. Scaffold-grade 2-by-10s can be used instead, but they must overlap both end frames by 6 to 12 inches. Both platforms have guardrails for safety, and optional toeboards *(page 31)* prevent workers from accidentally kicking bricks or other hazardous objects to the ground. A hoist lifts materials to the top of the scaffold.

### Scaffold Safety

✔ Be sure the scaffold is clear of overhead electrical wires. If power lines threaten to interfere with the placement of the scaffolding, ask the local power company if service can be temporarily shut off.
✔ Never use ladders or other makeshift devices to add to the height of the scaffold.
✔ Check all locking pins at end-frame joints and bracing attachments before climbing the scaffold.
✔ Keep both hands on the scaffold as you climb; put on a tool belt to carry tools.
✔ Never jump onto planks or platforms.
✔ Be sure the total load on the scaffold never exceeds the manufacturer's weight limits.
✔ Distribute the load on the platforms evenly.

## STABILIZING A SCAFFOLD

**Leveling the base.**

◆ When one level of scaffolding is together, position the assembly beneath the area where you will be working.
◆ Place a 2-by-10 plank lengthwise under each set of legs; on uneven ground, place individual 2-by-10 base pads under the legs.
◆ Hold a mason's level against the leg of one end frame while a helper turns the leveling jack until the leg is plumb *(right)*. If the jack cannot be raised high enough, add 2-by-10 pads under that leg.
◆ With the level still resting against the frame leg, adjust all of the other legs so that the scaffold is stable and remains plumb.
◆ Hold the level across the top of the end frame to check it for horizontal alignment; readjust the legs as necessary.
◆ Continue checking and adjusting until the scaffold is both level and plumb.
◆ Check it again each time you add a level, and adjust as necessary.

**Tying in a high scaffold.**

◆ At a height four times the narrower base dimension of the scaffold, mark the wall opposite the end frame at one end of the scaffold. For a masonry wall, drill a hole for a masonry anchor to fit a $\frac{1}{2}$-inch lag screw. For wood framing, drill a pilot hole for a $\frac{1}{2}$-inch lag screw long enough to penetrate a wall stud.

◆ Fasten a tie bracket to the wall with the lag screw.

◆ Attach a tie clamp to the end frame of the scaffold.

◆ With the wing nuts provided, join one end of the tie bar to the tie bracket and the other end to the clamp *(left)*.

◆ Repeat this procedure to tie the other end of the scaffold to the wall.

Where several frames are joined side by side, install an additional tie every 30 feet. Add another set of ties for every 26 feet that the scaffolding rises above the first set.

## TOEBOARDS FOR SAFETY

A brick falling from the materials platform can present a potentially fatal danger to workers below. Toeboards, available through the scaffold rental agency, reduce the chance of an accident by providing a fence around the edge of the materials platform. The toeboards, usually metal lengths with interlocking metal plates at each end, are positioned on the inside of the guardrail posts.

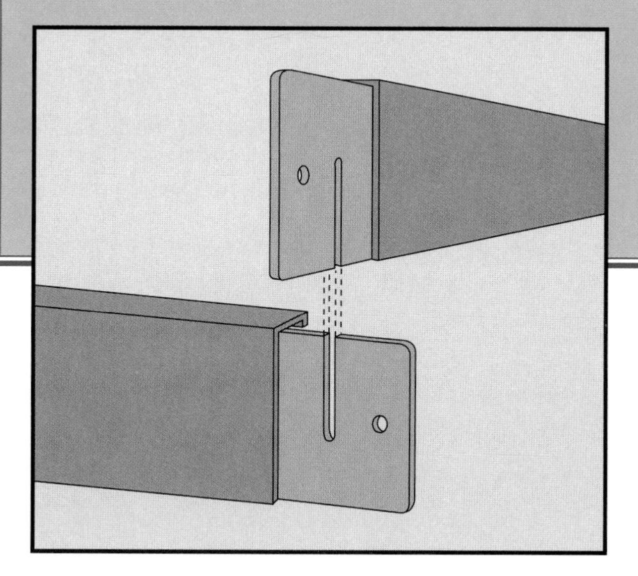

# Building with Concrete

Thorough preparation before pouring a concrete footing or slab is an essential part of achieving a strong foundation. In addition to laying out the building plan, the job may entail rough-grading the area, constructing forms, and adding steel reinforcement. Once the site is ready, you can pour the concrete. Concrete can also be cast into blocks or combined with soil to make adobe bricks.



Smoothing a decorative block →

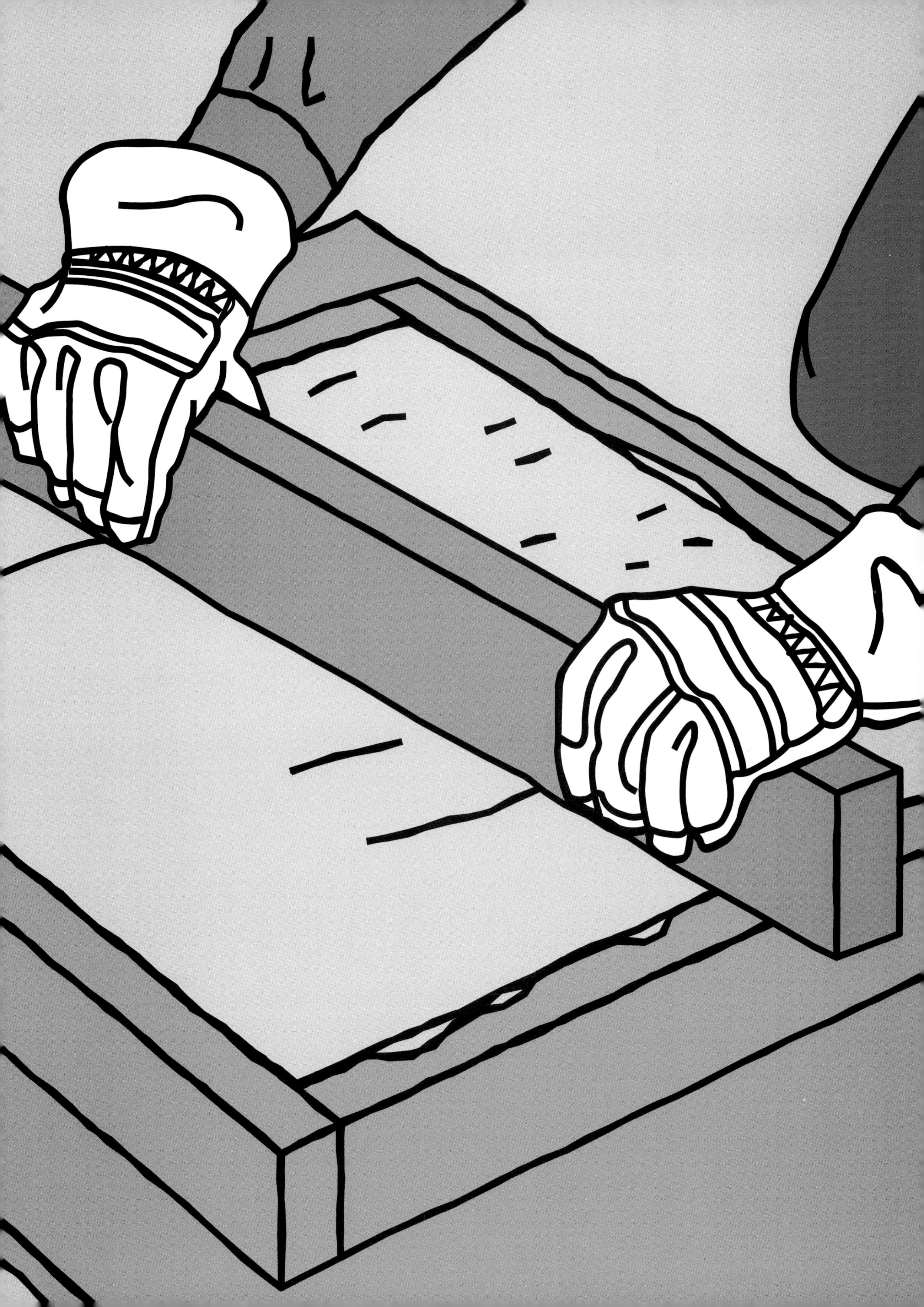

# Marking Building Lines

Careful site preparation will not only save you time, but may forestall structural or even legal complications. For example, the dimensions of most footings and slabs take advantage of modular building materials such as standard concrete blocks; if the site is poorly measured, the modules may have to be custom-cut, And if a building line is laid slapdash, it may come closer to property lines than the law allows.

**Locating the Property Line:** In many cases, steel stakes mark the corners of the lot, and you can find them with a metal detector. If not, you will need the property map—the plat—on file with the county zoning office or accompanying the property deed; the plat shows permanent fixtures that were on hand when it was surveyed—utility poles, house walls, and the like—to which you can refer to locate boundaries. The measurements may not be exact, however; if there is a chance that the new structure will violate setback laws or transgress a property line, have the land resurveyed and marked with corner stakes.

**Laying Out the Site:** If the terrain is uneven, do some rough-grading first *(page 42)*. For a small project on fairly level land, you can lay out the project with the 3-4-5 method on the following pages; but when the site is sloping or hilly or the project is large, it may be impossible to lay out straight, accurate lines with this method. Here you will need the greater precision of a laser level *(pages 38-40)*; the tool and its accessories can be rented from a surveying-equipment supplier.

CAUTION *Before positioning or excavating trenches, find the locations of underground obstacles such as dry wells, septic tanks, and cesspools, and electric, water, and sewer lines.*

**TOOLS**

- Maul
- Hammer
- Two long tape measures
- Sledgehammer
- Water level
- Handsaw
- Carpenter's level
- Plumb bob
- Laser level, tripod, and leveling rod
- Beam splitter

**MATERIALS**

- Stakes
- String
- Common nails (2")
- Double-headed nails (2")
- 1 x 6s, 2 x 4s
- Powdered chalk

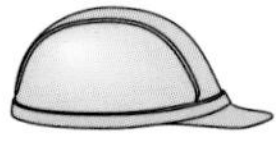

**SAFETY TIPS**

*Wear goggles when hammering or driving stakes.*

## LAYING OUT A SITE WITH THE 3-4-5 METHOD

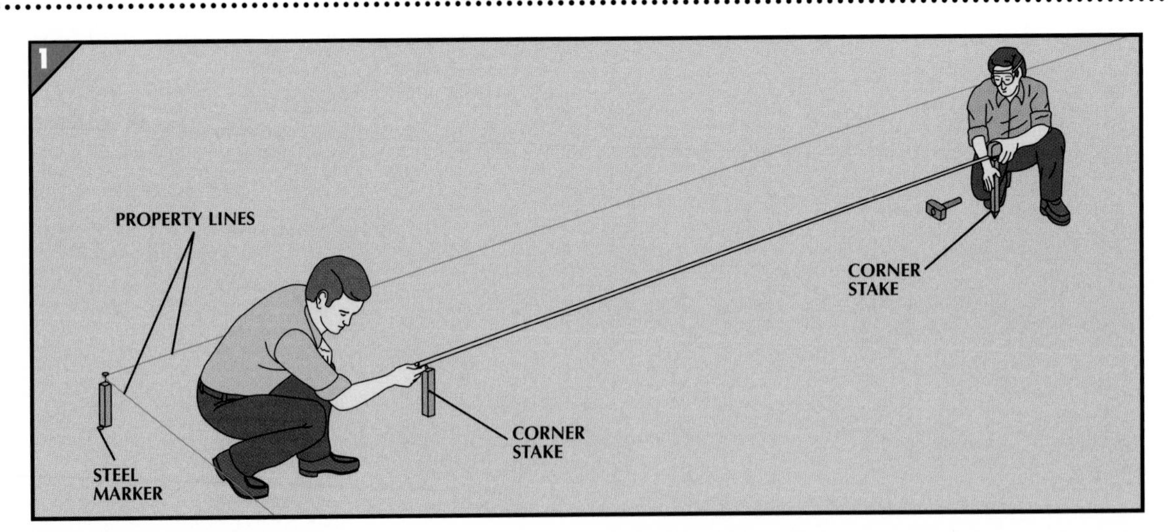


**1. Setting the first building line.**

◆ Sink stakes just inside the steel property-line markers; drive nails into the tops of the stakes and run strings between them.

◆ Starting at the corner of the planned project nearest the property lines, drive a stake to mark the first corner.

◆ With a helper, measure off the proposed length of one side of the planned building and drive a stake for the second corner *(above)*.

◆ Drive nails into the tops of the two stakes, then tie a string between the nails, allowing an extra 5 feet of string to trail from each stake.

◆ Measure between the stretched string and the property line to be sure the building is far enough from the line.

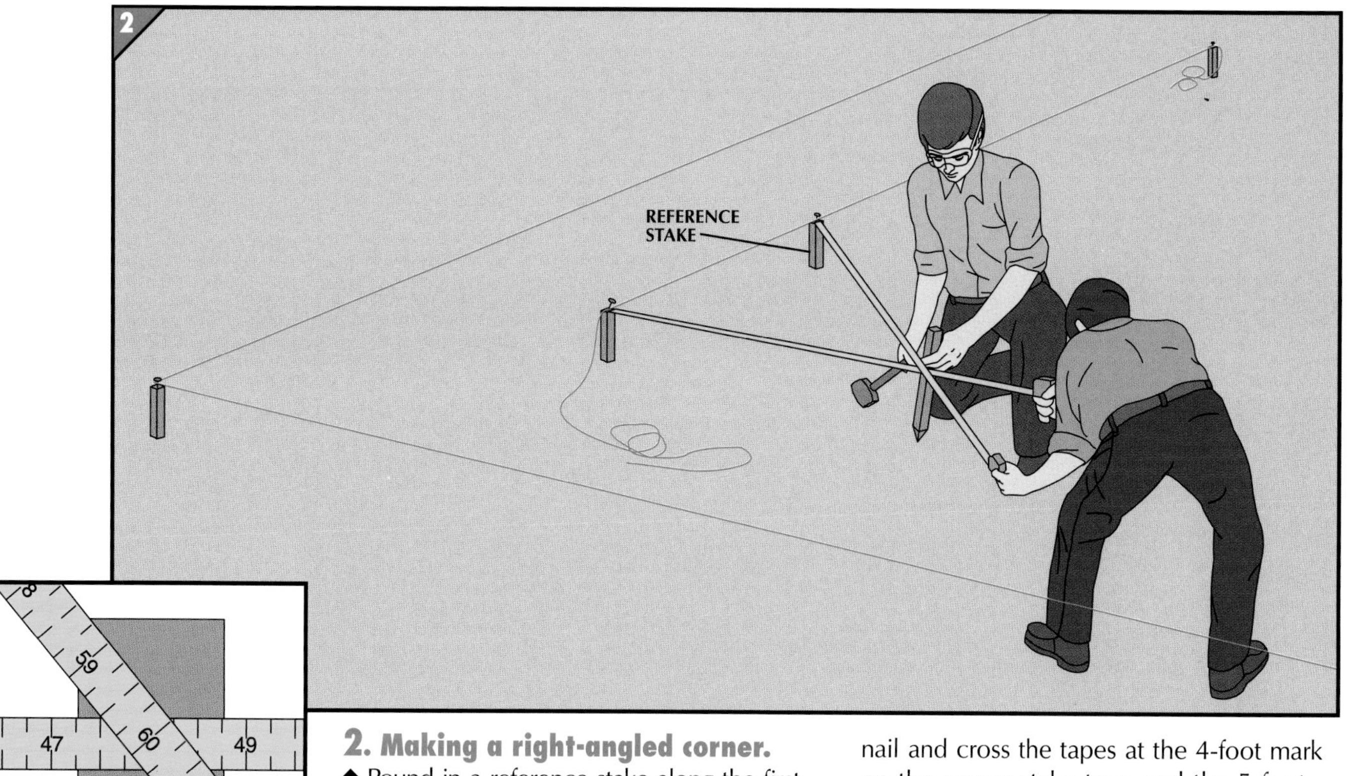


## 2. Making a right-angled corner.

◆ Pound in a reference stake along the first building line so that its center is exactly 3 feet from the nail in the first corner stake, and drive a nail into the center of it.

◆ Hook a long tape measure over each nail and cross the tapes at the 4-foot mark on the corner-stake tape and the 5-foot mark on the other tape *(inset)*.

◆ Have a helper drive a second reference stake at the point at which the two tape measures intersect *(above)*.

## 3. Finishing the lines.

◆ From the corner stake, stretch a tape measure the planned length of the second building line.

◆ With your helper sighting over the second reference stake, center the tape over this stake; center a second corner stake under the end of the planned building line *(above)*, then drive it in.

◆ Hammer a nail into the top of the second corner stake; tie a string between the two nails, with an extra 5 feet at each end.

◆ Double-check all the measurements.

◆ Stake right angles at the remaining corners to complete the other two building lines.

◆ Check that the building lines are far enough away from property lines, then remove the property line strings and the reference stakes, leaving the four corner stakes and the building line strings in place.

# SETTING UP BATTER BOARDS

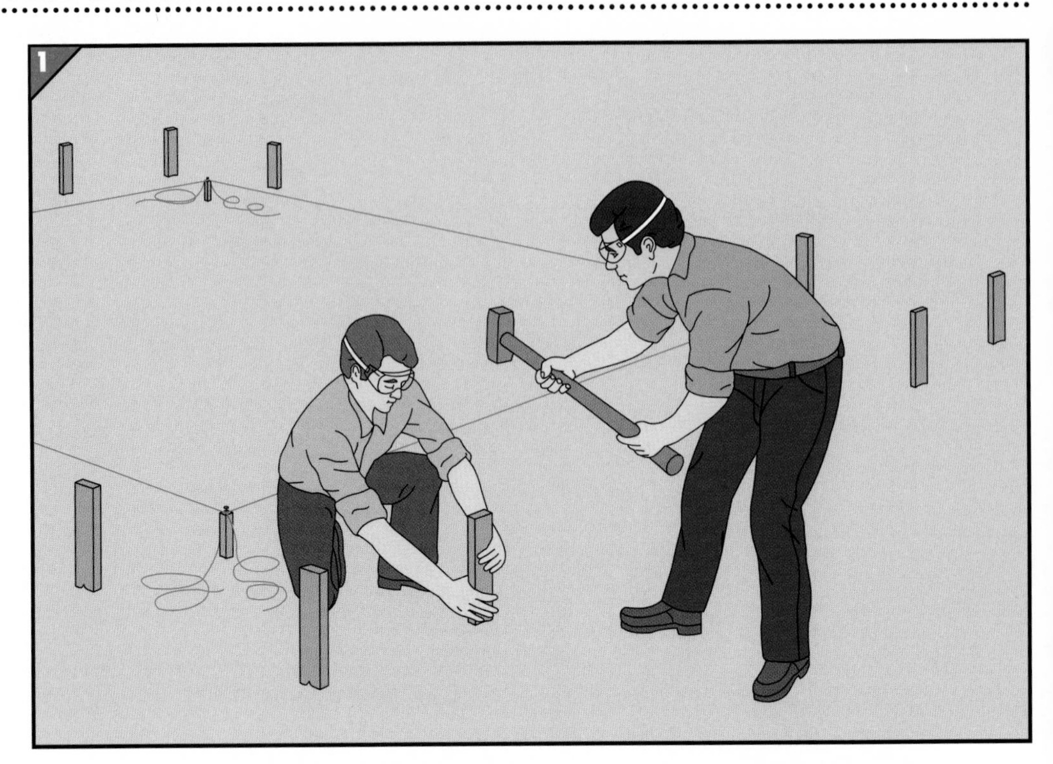


**1. Placing the batter-board stakes.**
At each of the four corner stakes, parallel to the string lines and about 2 feet outside the building lines, drive in three 2-by-4 stakes with a sledgehammer *(right).*

**2. Marking level lines.**

◆ On the outside surface of one 2-by-4 corner stake—the one at the highest corner if the site is not completely level—draw an elevation line about 10 inches above the ground.

◆ Make a water level by filling a length of flexible transparent tubing with water and food coloring—or buy a commercial model.

◆ With a helper holding one end of the water level against the stake at the marked elevation line, stretch the tube to one of the other corner stakes, adjusting the height of the hose until your helper signals that the water is level with the line. Mark this height on the stake *(above).*

◆ Mark this level on the two remaining corner stakes in the same manner.

## 3. Fastening the batter boards.

◆ Cut eight 1-by-6s, each long enough to run from the outer stakes to the corner stakes.

◆ Start pairs of 2-inch double-headed nails through the boards at the points where they will attach to the stakes.

◆ Place one board with its end flush with the corner stake and its top edge even with the marked line; drive one nail into the stake.

◆ Rest a carpenter's level on the board and pivot the board on its nail until it is level *(right)*.

◆ Secure the board to the outer stake with a single nail, then drive the remaining two nails.

◆ Using the same technique, attach a second batter board between the corner stake and the second outer stake.

◆ Fasten pairs of batter boards at the remaining three corners.

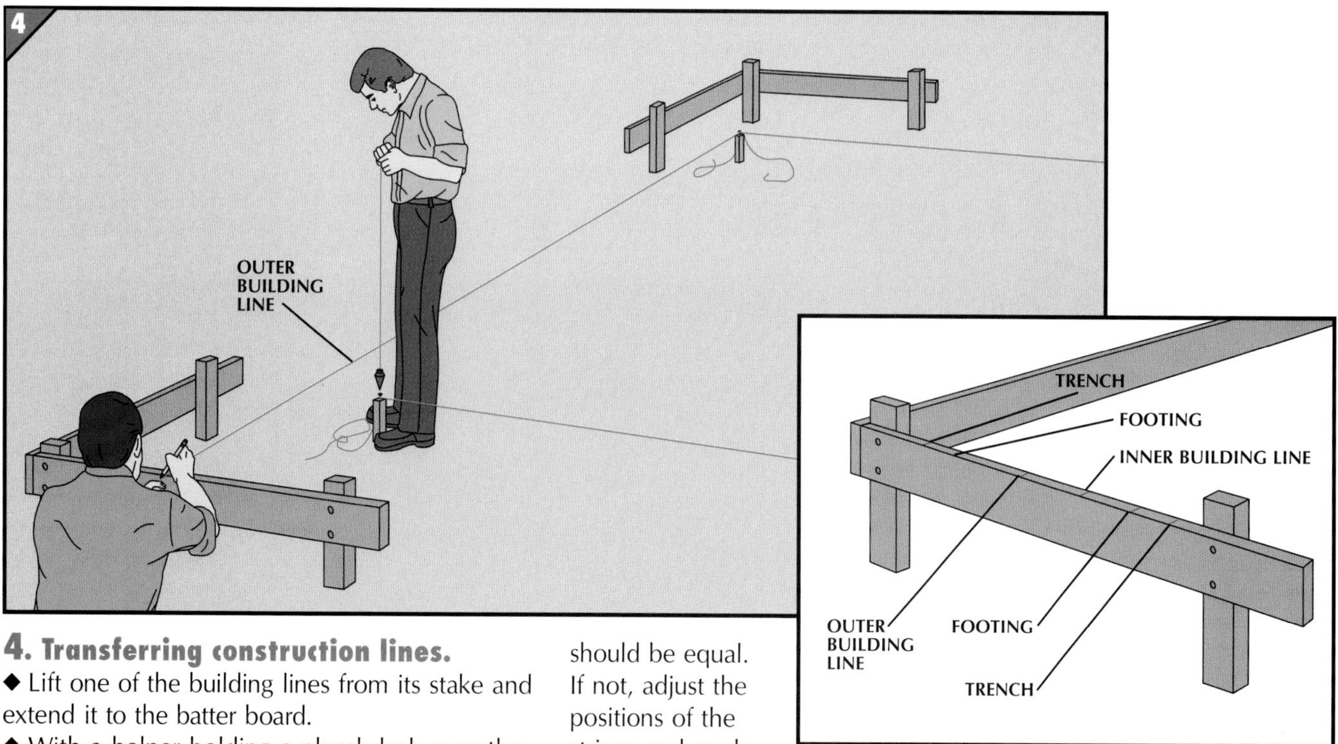


## 4. Transferring construction lines.

◆ Lift one of the building lines from its stake and extend it to the batter board.

◆ With a helper holding a plumb bob over the nail on the original corner stake, align the building-line string with the plumb-bob string.

◆ Tie the string around the board, then make a mark at the string, defining the position of the outside of the building wall *(above)*. Repeat the procedure to mark the outside of the wall on all the batter boards.

◆ Check that the layout is square by measuring the diagonals between opposite corners—they should be equal. If not, adjust the positions of the strings and marks on the batter boards, making sure that the length of the building lines remains correct, then remove the strings.

◆ As shown in the inset, use these marks to measure and mark the positions of the inside of the building wall *(page 70)*, and the footing *(pages 45-46)* on the batter boards. Anchor the strings on the batter boards by cutting a groove at each mark with a handsaw.

**5. Liming the lines.**
◆ Stretch string lines between the grooves on the batter boards for the footing—or, in sandy soil, between the grooves for the trench.
◆ Fill a plastic squeeze bottle with powdered chalk, and trim the bottle tip at an angle to produce a fairly wide line.
◆ Working by eye, trickle a line of chalk on the ground directly beneath each of the string lines from one end to the other—or have a helper move ahead of you with a plumb-bob string against the string line while you dispense the chalk.
◆ Remove the strings.

## SIGHTING WITH A LASER LEVEL

**1. Assembling and leveling.**
◆ Set up the level's tripod, lock the level in the mounting clamp, and hang a plumb bob from the hook beneath the swivel base.
◆ Drive a stake just inside the property-line corner marker that is closest to the planned building and move the tripod until the plumb bob hangs over the stake's center.
◆ Turn the swivel base to position the level vial parallel to two of the leveling feet; adjust the feet to center the bubble in the vial *(right)*.
◆ Rotate the swivel base 90 degrees in either direction and adjust the third leveling foot to center the bubble in the vial.
◆ Continue to turn the swivel base, adjusting the feet until the bubble stays centered when the base is in any position.
◆ Adjust the vertical micrometer until the bubble in the level vial on the laser level itself is centered.
◆ Swivel the laser 180 degrees and check the bubble. If it is slightly off center, adjust the vertical micrometer to bring the bubble halfway to the center of the vial, then adjust the third leveling foot to center the bubble completely.

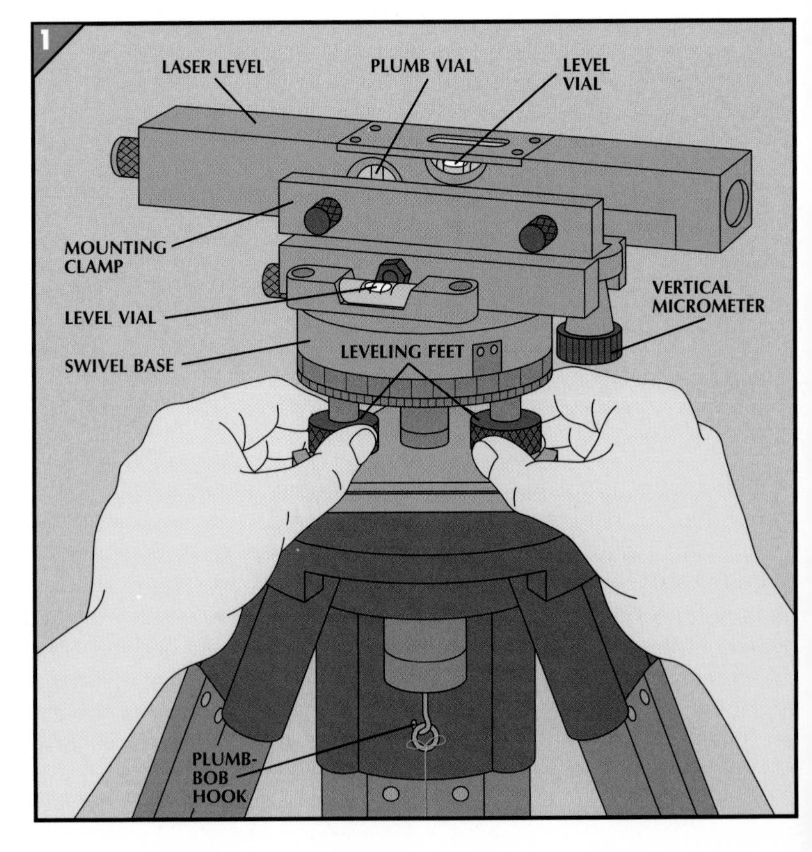

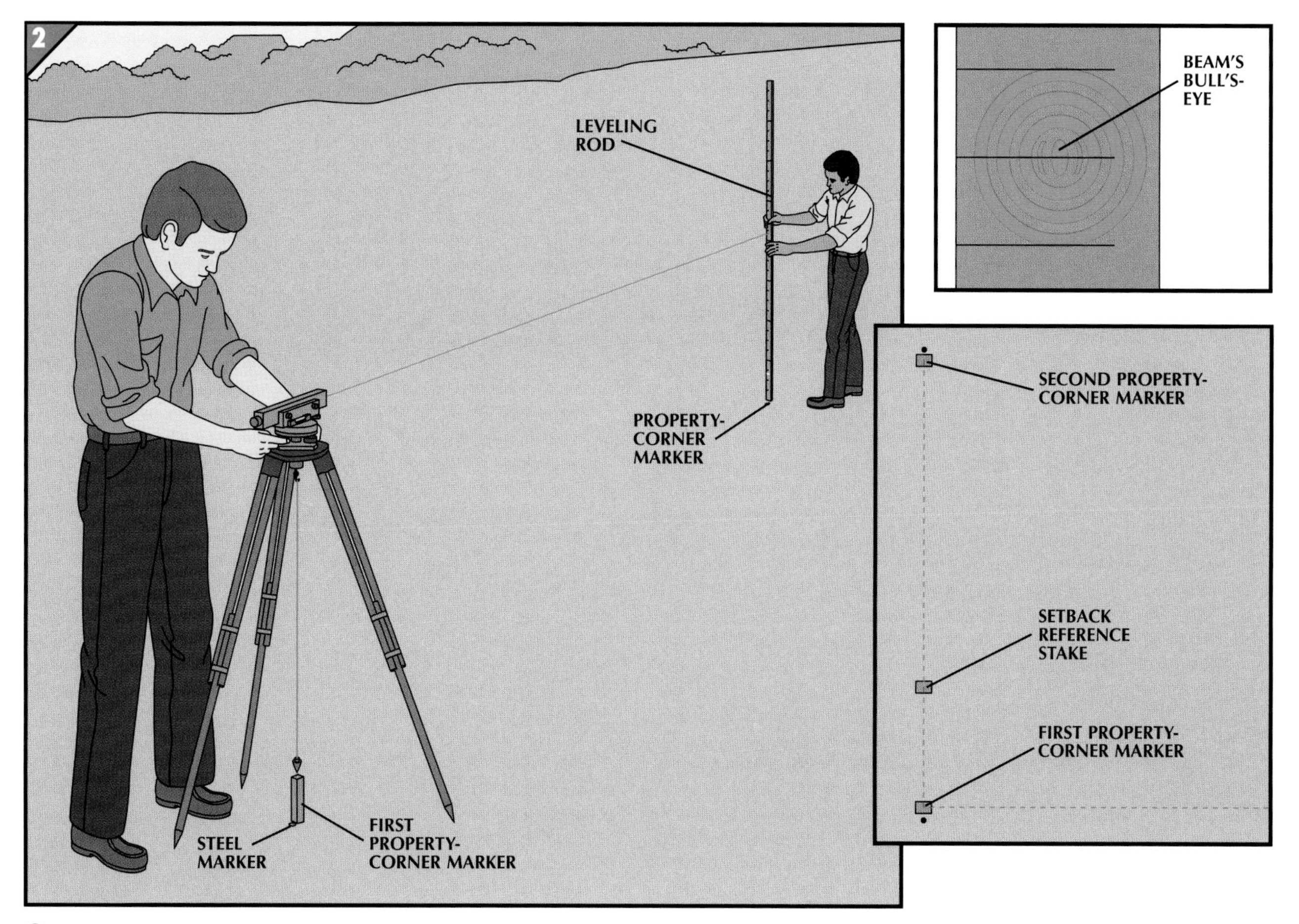


**2. Plotting the setback.**

◆ Drive a stake into the ground in front of the second property-corner marker, and have a helper hold the leveling rod directly over the stake.

◆ Fire the laser beam in the direction of the rod, then move the swivel base side to side *(above)* until your helper indicates that the beam's bull's-eye is centered on the rod, as shown in the upper inset.

◆ Have your helper move the leveling rod along the property line to the point of the planned setback.

◆ Without moving the level, have him align the rod with the bull's-eye.

◆ Drive a setback reference stake at the rod, aligned with the property line, as illustrated in the lower inset.

## A BEAM SPLITTER FOR RIGHT ANGLES

When attached to a laser level, a beam splitter—by means of two built-in mirrors—produces a second beam at a 90-degree angle to the level's beam. To use the splitter, slide it over the lens of the laser level and secure it in place with the locking screw. When the laser is turned on, the main beam exits the front aperture of the splitter, and the second beam exits the side lens. You can then use the two beams to plot the corners of a building *(page 40, Step 3)*.

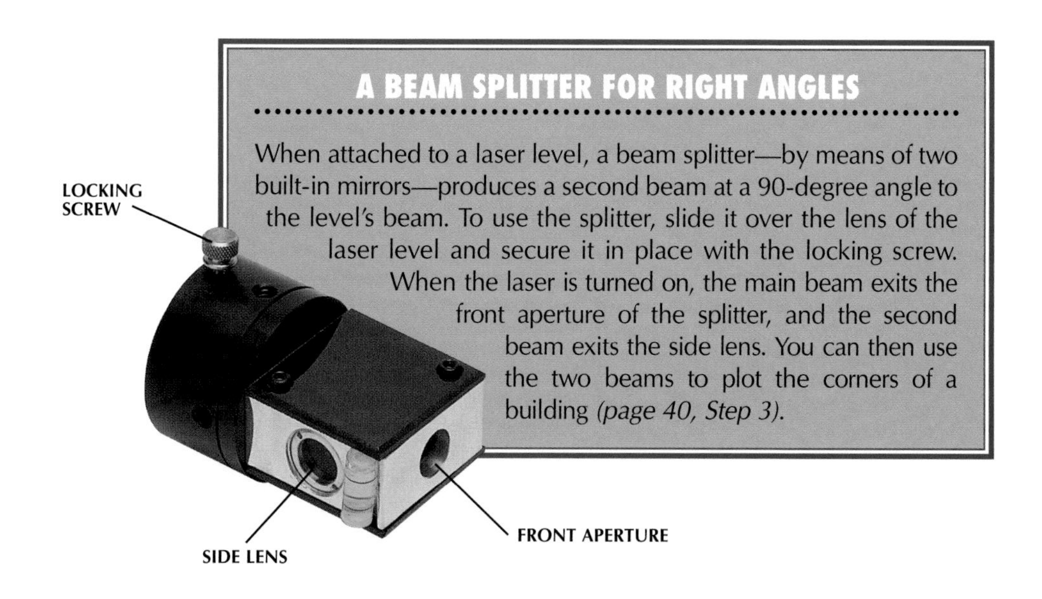

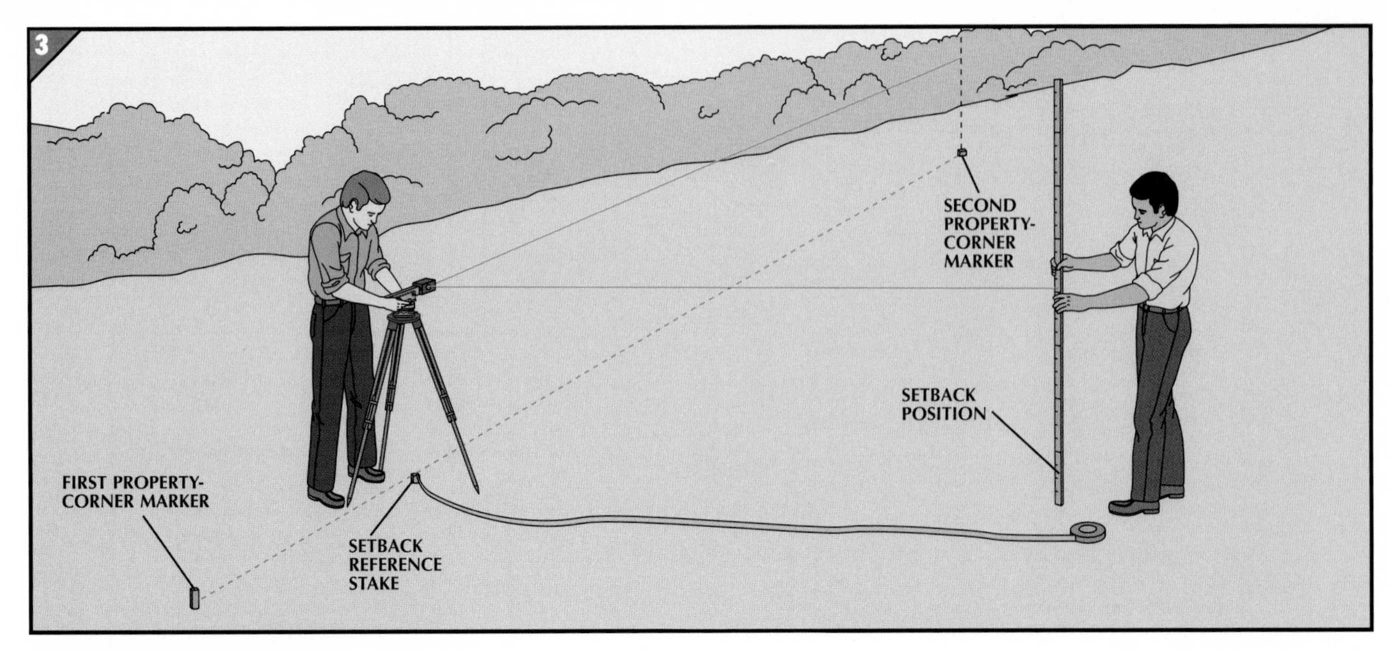


### 3. Marking the building's corners.

◆ Attach a beam splitter *(box, page 39)* to the level, and set it on the property line a few inches beyond the setback reference stake.
◆ With the red dot from the beam splitter aiming straight down, shift the entire unit to center the dot over the setback reference stake.
◆ Level the laser, then adjust the position of the beam splitter so both beams shoot out parallel to the ground.
◆ With your helper holding the leveling rod at the second property-corner marker, center the bull's-eye on the rod.
◆ Have your helper move the leveling rod to line up with the second beam from the splitter *(above)*. Place the rod so that it is the required setback from the property line, centering the rod on the beam. Drive a stake at this point to mark the first corner of the building.
◆ Without moving the level, set a second stake farther along the same sight line to mark the second corner of the building.
◆ Place the level at either building-line corner, level it, then using the same technique for marking the setback line, sight a 90-degree angle to stake an adjacent building line.
◆ Remove the splitter from the laser.
◆ Measure the dimensions of the layout to be sure they are correct.

### 4. Leveling the batter boards.

◆ Drive stakes for batter boards at all four corners of the building line *(page 36, Step 1)*. At the highest corner of the site, mark one stake about 10 inches above the ground.
◆ Set up and level the laser level near the center of the site.
◆ Have a helper hold the leveling rod against the marked stake, lining up the bottom of the rod with the mark.
◆ Fire the laser and have your helper read the number where the bull's-eye hits the rod *(left)*; record this number.
◆ Move the leveling rod to another corner stake, and swing the laser's swivel base to focus on the rod in that position.
◆ Have your helper move the rod until the beam hits the recorded number, then mark the stake at the bottom of the leveling rod.
◆ Repeat this procedure to mark the remaining two stakes.
◆ Install batter boards and put up construction lines *(pages 37-38, Steps 3-5)*.

# Measuring for a Stepped Footing

When a lot slopes steeply, you can save on excavation and materials with a footing that descends in a series of steps. To erect a concrete-block wall on the footing, the steps' risers must be in multiples of 8 inches—the height of one course of blocks—so that the blocks will rise level with each tier of steps. The exact length of the steps is not crucial, but for structural reasons make them at least 2 feet long.

To lay out this footing, use the techniques on pages 38-40, setting up the batter boards at the corners. Once you have marked the elevation changes, the excavator will use the marks to dig to each different depth.

**CAUTION** *Before positioning or excavating trenches, locate underground obstacles such as dry wells, septic tanks, and cesspools, and electric, water, and sewer lines.*

**1. Measuring grade changes.**

◆ Tie strings between batter boards along the inner or outer building lines. If you will be using a laser level *(pages 38-40)*, set it up at the approximate center of the site, and level it *(page 38, Step 1)*.

◆ Beginning at the downhill end of one string and gradually working uphill, measure the distance between the string and the ground with a leveling rod *(above)*. With a laser level, have a helper stand a leveling rod on the ground next to the lowest corner stake, then swivel the level until the laser beam hits the leveling rod; note the number where the bull's-eye hits. Slowly move the beam along the length of the line while your helper walks uphill, keeping the beam on the leveling rod.

◆ Drive a stake wherever the level changes 8 inches—or 24 inches on a steeper slope—then write "8" or "24" on each stake.

**2. Marking step locations.**

◆ Starting at a corner and working uphill along one side of the site, measure the horizontal distance between each stake marking the grade change.

◆ Wherever the distance is less than 2 feet, remove the uphill stake and mark the next stake "16" or "48" to indicate to the excavator that it represents a 16- or 48-inch grade change, rather than one of 8 or 24 inches.

◆ Repeat the measuring, staking, and marking procedure on any other sloped sides.

◆ With a plastic bottle filled with chalk, mark a perpendicular grade line at each stake marking a rise in grade, extending the lines 2 feet beyond each side of the footing or trench lines *(left)*.

# Rough-Grading a Building Site

Some lots require preliminary leveling, or grading, to provide a relatively even working surface *(opposite)*. The boundaries of the rough-graded area are up to you to decide, and need not be laid out as precisely as the perimeter of the planned structure. Establish the level of the grade in reference to a known elevation—usually indicated on builder's plans. You will then need to indicate on the site the amount of earth to be cut or filled.

Before beginning the rough-grading procedure, clear the building area of trees, stumps, and brush.

**Planning the Excavation:** Prior to digging, you need to know what to expect below the surface. If you plan to go deeper than the soft, workable layer of topsoil, have an engineer take a soil bore or a core sample; these tests will indicate if there is a subsurface of stony soil or solid rock.

Hire a professional to do the actual excavating, to ensure that it is done safely and accurately. The fee will be determined by the size of the cut, the rock content of the soil (excavating in stone is much more expensive) the distance spoil—excavated earth—must be hauled, and the time required for digging and shoring.

**Preserving Earth:** If you want to save topsoil scraped off during grading, select a spot for it to be piled, away from the excavation and construction area. If you have earth hauled away, be sure to save some spoil for backfilling the hole once the structure is finished.

CAUTION *Before positioning or excavating trenches, find the locations of underground obstacles such as dry wells, septic tanks, and cesspools, and electric, water, and sewer lines.*

TOOLS

Long tape measure
Laser level, tripod, and leveling rod
Plumb bob
Sledgehammer

MATERIALS

Stake
2 x 4s

## TRICKS OF THE TRADE

### Protecting Tree Trunks

To save trees from damage by grading and excavation equipment, pad them with several layers of thick cloth or newspaper to a height of 4 feet, then surround the padding with a paling of closely spaced 1-by-4 boards; fasten it at top and bottom with baling wire *(right)*. Paint the roots white with latex paint where they enter the ground; should the base of the trunk be temporarily buried, refer to the painted area to keep from digging too close and damaging the roots.

# MARKING CHANGES IN ELEVATION

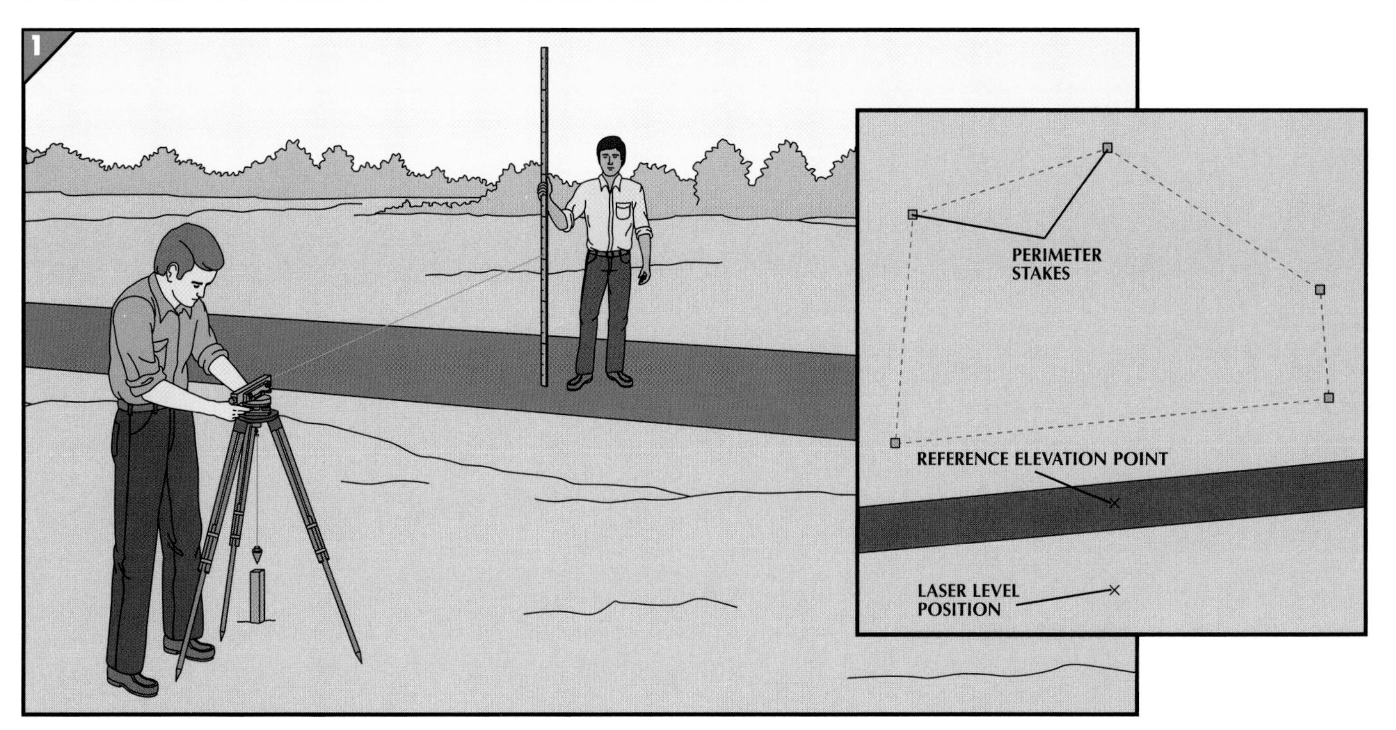


**1. Setting the grade stakes.**

◆ Cut 2-by-4 stakes that will extend 3 feet out of the ground when driven into it. Drive the stakes at 30-foot intervals, laying out the perimeter of the area to be graded.

◆ To fix the spot from which you will be sighting, drive a stake outside the area on its lowest side, and set up and level a laser level over it *(page 38, Step 1)*. As shown in the inset, align it with the reference-elevation point—in this example, a road—which is noted on the blueprint or building plan.

◆ Have a helper hold a leveling rod on the ground at the reference-elevation point, then fire the beam at the rod and record the reading *(above)*.

◆ Calculate the difference between this reading and the rough-grade level indicated on the plan, keyed to the reference elevation. Make a note of this difference; it will be used as a reference reading.

◆ With your helper holding the leveling rod on the ground at one of the perimeter stakes, pivot the laser to focus on the rod.

◆ Have your helper note the difference between this reading and the reference reading, indicating with a plus or minus whether it is higher or lower. Transfer this figure to the grading plan.

◆ Without repositioning the tripod, pivot the laser in this way to find the difference in the readings for each perimeter stake.

**2. Marking the stakes.**

◆ With a felt-tipped marker, draw a line at ground level on each perimeter stake, and above the line draw an arrow pointing downward to help locate the line as the earth is moved.

◆ Label each stake with a "C" if the earth is to be cut away at that point, or an "F" if the excavator will need to fill there instead.

◆ Using the plus and minus figures noted on the grading plan, on each stake mark the amount of earth to be cut or filled at that point, making the figures large and clear enough for the excavator to read them from the machine *(left)*.

# Constructing a Footing

The footing of a structure carries weight down to the solid, unexcavated soil below the frost line. It provides a firm, flat surface on which to build and, since it is typically twice as wide as the wall, distributes the load over a greater area of soil. Dimensions of footings are determined by codes; a residential building's footing is generally 8 inches thick; those for piers may be up to 3 feet thick.

**Preparing the Trench:** In compact soil with a lot of clay, the trench for the footing can serve as a form for the concrete. Grade pegs of steel reinforcing bar (rebar) are set into the earth so their tops are level with the planned top of the footing. When soil is unstable, wooden forms shape the concrete, and their tops indicate the footing height. Forms are made of 2-inch-thick lumber as wide as the footing is thick; for an 8-inch footing, for example, you'll need 2-by-8 boards. Because a 2-by-8 is only $7\frac{1}{2}$ inches wide, raise the forms $\frac{1}{2}$ inch and pack the gap under them with earth before you pour the concrete. Curved footings generally use plywood *(page 46)*. Most codes require that footings be strengthened by rebar *(pages 47-49)*.

**Stepped Footings:** A footing on a steep slope *(page 49)* is built with baffles—form boards that shape the concrete at the front of each step. The bottom edge of the first baffle should be the same distance above the trench floor as the planned thickness of the footing. Because the steps will overlap about 2 feet, have the excavator dig each step about 2 feet forward of the original step markers.

**Concrete:** Multiply the length, width, and thickness of the footing to get its volume in cubic feet; then convert this into cubic yards—the form in which concrete is sold—by dividing it by 27. A dry, sunny day will speed up the setting of the concrete. In this case, have a third person working with you, following behind and smoothing the surface *(page 50)* while you and your helper pour and spread it.

**CAUTION** *Before positioning or excavating trenches, find the locations of underground obstacles such as dry wells, septic tanks, and cesspools, and electric, water, and sewer lines.*

**TOOLS**

- Carpenter's level
- Tape measure
- Water level
- Rebar cutter
- Maul
- Tamper
- Plumb bob
- Hammer
- Handsaw
- Circular saw
- Rebar bender
- Wheelbarrow
- Square shovel
- Float
- Pliers
- Hand spade

**MATERIALS**

- 1 x 4s
- 2 x 2s, 2 x 4, 2 x 8s
- Plywood ($\frac{1}{4}$" or $\frac{1}{8}$", $\frac{1}{2}$")
- Double-headed nails (2", 3")
- Rebar
- Tie wire
- Chalk
- Concrete pieces or rebar supports
- Concrete
- Plastic sheeting

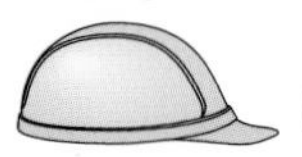

**SAFETY TIPS**

*Protect your eyes with goggles when nailing; add gloves when bending rebar, and rubber boots when pouring or spreading concrete.*

## CONCRETE FROM A TRUCK

The most convenient way to buy large amounts of concrete is from a transit-mix company that will mix it to your specifications and deliver it ready to pour.

Most footings call for air-entrained concrete: a mixture that contains 4 percent air is usually sufficient, but the supplier will know the right kind for your region. Stepped footings need a stiff mixture that will set up quickly.

Just before the truck arrives, hose down the inside of the forms to prevent them from drawing water from the concrete. If possible, have the truck back right up to the footing trench. Otherwise, lay down a ramp of 2-by-8s between the truck and the trench, and enlist several helpers to transport the concrete in wheelbarrows.

# PREPARING A PACKED-EARTH FORM

**Setting grade pegs.**

◆ Drive a 2-by-2 stake at each corner of the trench, then 3 inches in from either side of the trench drive stakes every 3 feet, alternating sides.

◆ Check the trench bottom for level with a carpenter's level and shave down any high points where necessary.

◆ Mark the desired thickness of the footing—in this case, 8 inches—on the reference stake *(right)*, then transfer this grade mark to the other stakes with a water level *(page 36, Step 2)*.

◆ Drive a grade peg—a short length of rebar—into the ground beside each stake, until its top is level with the grade mark *(inset)*.

◆ Remove the stakes and tamp the earth firmly around the grade pegs.

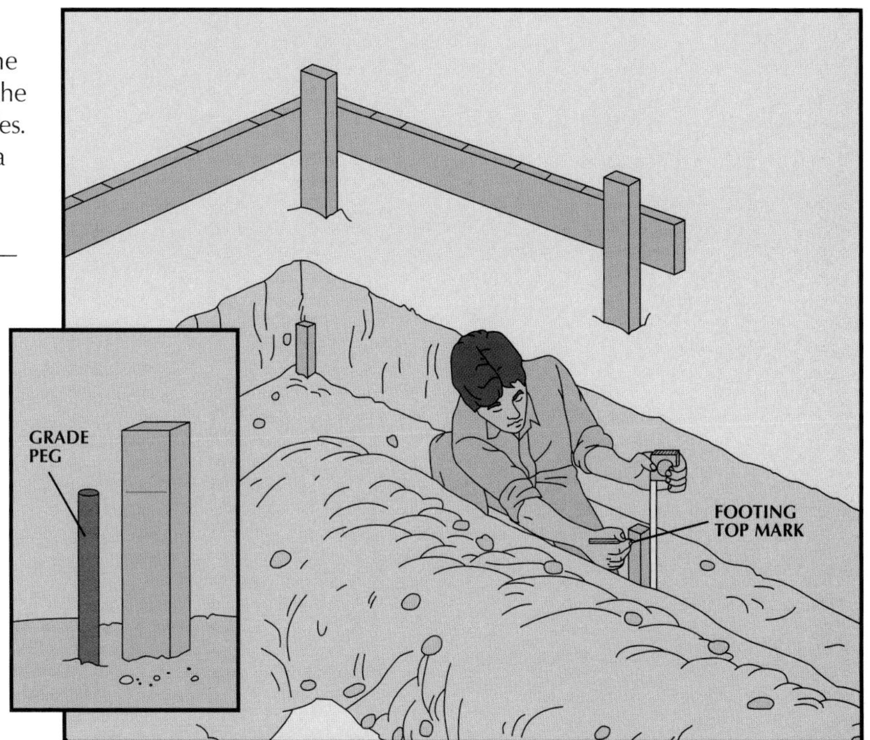


# INSTALLING FORM BOARDS

**1. Setting up the outside form boards.**

◆ Put up string lines on the batter boards to mark the outside edge of the footing.

◆ Drop a plumb bob where the strings intersect. At a point $1\frac{1}{2}$ inches behind the plumb bob, drive in a 2-by-2 support stake, then drive another stake at a right angle to the first and the same distance from the bob *(above, left)*.

◆ Drive stakes at the other corners in the same way.

◆ Sink stakes at 3-foot intervals along each trench, holding the plumb bob against the string line to position them.

◆ Mark the top of the footing on a reference stake and transfer it to the other stakes as you would for a packed-earth footing.

◆ Cut form boards from 2-inch lumber as wide as the footing is thick.

◆ With 3-inch double-headed nails, attach form boards to the insides of the stakes, level with the footing marks; back the boards with a maul when nailing stakes to boards *(above, right)*. Set an extra support stake where any two form boards join.

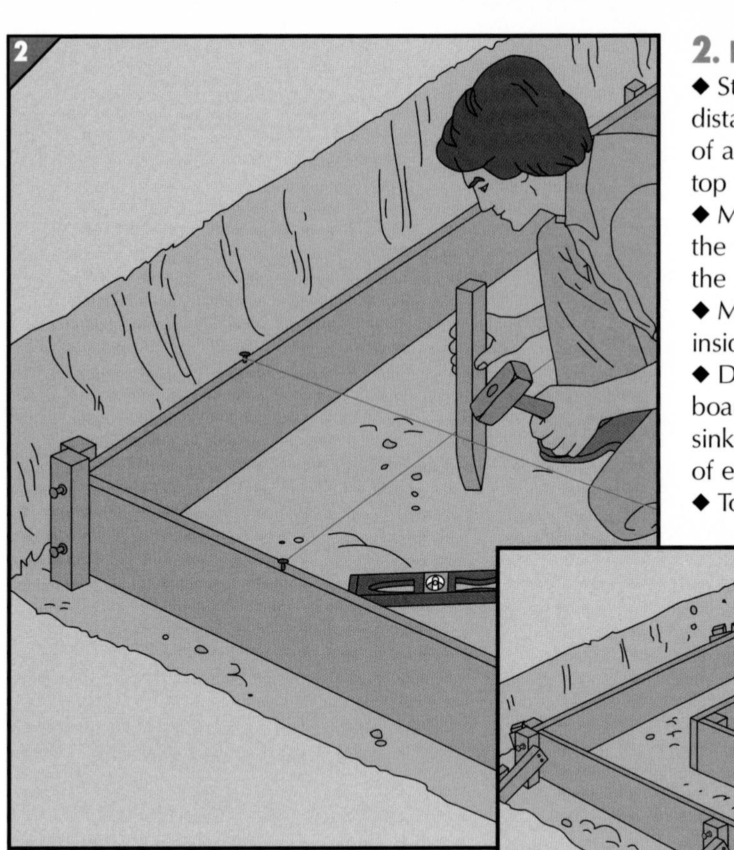


**2. Building the inside forms.**

◆ Starting at one end of an outside form board, measure a distance equal to the width of the footing plus the thickness of a form board; mark this point by driving a nail into the top of the board.

◆ Measure and mark the form board at the opposite end of the trench at the same distance, then run a string between the nails to mark the position of the inside form board.

◆ Measure, mark, and add string lines for the rest of the inside form boards in the same way.

◆ Drive a pair of 2-by-2 support stakes for the inside form boards at each corner where the strings intersect *(left)* and sink individual stakes at 3-foot intervals down the length of each string.

◆ To mark the footing level on each stake, rest a carpenter's level on the top of the outside form board, and holding it level, mark the bottom of the level on the stake.

◆ Nail form boards to the insides of the stakes *(Step 1)*, then cut off all the stakes flush with the top of the form boards.

◆ Brace the support stakes with 1-by-4s anchored to outrigger stakes as shown in the inset.

◆ Pack earth halfway up the outside of the forms.

# BUILDING A CURVED FORM

**Setting out a curve.**

◆ Follow the trench curves to mark two parallel chalk lines the width of the footing in the trench.

◆ Along the inner curve line, drive 2-by-2 support stakes 1 foot apart and $\frac{1}{4}$ inch outside the line. On the inside of the outer curve, drive temporary stakes 1 foot apart and directly at the line.

◆ Mark the top of the footing on a reference stake and transfer it to the other stakes as you would for a packed-earth footing *(page 45)*.

◆ Cut a form the height of the footing from $\frac{1}{4}$-inch plywood—or two thicknesses of $\frac{1}{8}$-inch plywood for tight curves.

◆ At the inner curve stakes, fasten the plywood to the stakes with 3-inch double-headed nails with its top even with the footing-level marks.

◆ Nail a form to the temporary outer-curve stakes, then drive stakes outside the curve, against the plywood and staggered from the inner, temporary stakes *(right)*. Nail the plywood to the outer stakes, then remove the temporary ones.

◆ Cut off all the stakes flush with the plywood.

◆ Brace each stake with an outrigger stake as for a standard form board and push earth halfway up the form to brace it.

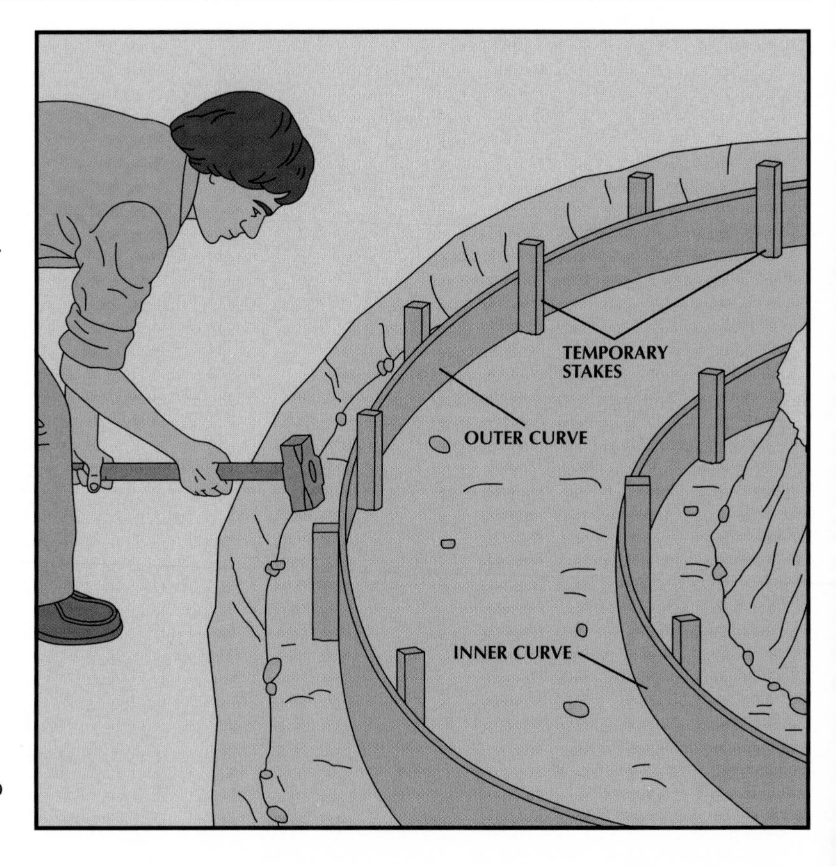

# FRAMING A STEPPED FOOTING

### Stacking the layers.

◆ To build a three-sided frame for the first step, make form boards as wide as the footing is thick. To make 8-inch form boards, cut strips of $\frac{1}{2}$-inch plywood $1\frac{1}{2}$ inches wide and tack them to the bottom of 2-by-8s. Cut one form board to the width of the footing and two to the length of the lowest step, then fasten them together with 3-inch double-headed nails.

◆ Position the frame in the trench with the open end against the first earthen step.

◆ Drive two 2-by-2 stakes a few inches back from the front end of the frame, and two taller stakes about 20 inches in front of the earthen step.

◆ Attach the frame to the stakes with 3-inch double-headed nails, holding a maul behind the form boards when you are nailing.

◆ Drive two stakes near the back of the second step, footing-width plus 3 inches apart.

◆ Cut two form boards 2 feet longer than the length of the next step.

◆ Position the boards on the second step with their ends against the third step, and on top of the boards on the first step. Nail all four stakes to the form boards.

◆ Continue to install form boards on each step in this way.

◆ Lay rebar to reinforce the footing *(page 49, top)*.

◆ Add a form board at the front end of all the other levels.

◆ Cut the tops of the stakes flush with the boards.

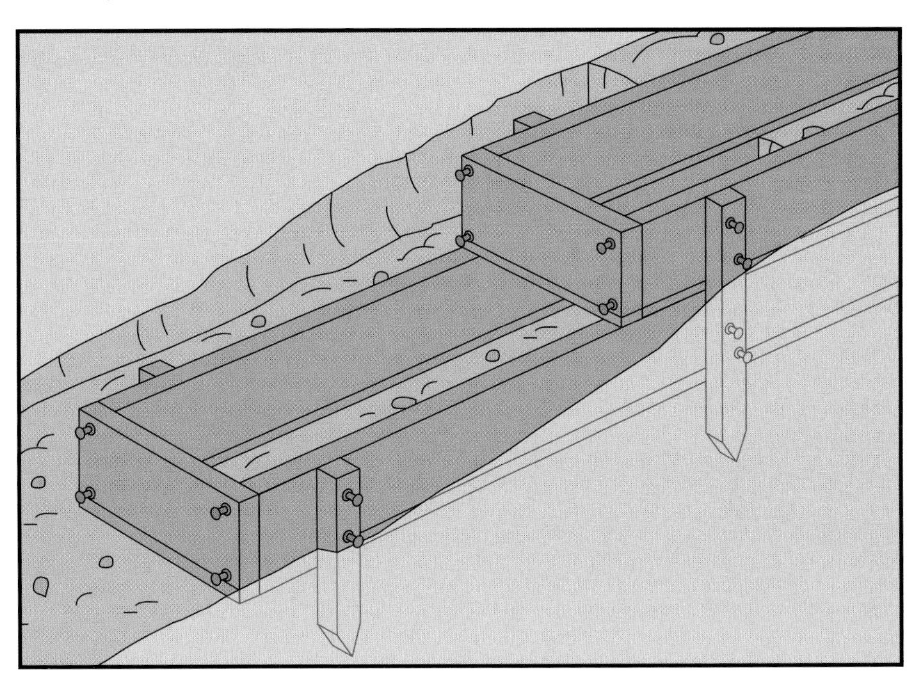

# ADDING REINFORCEMENT

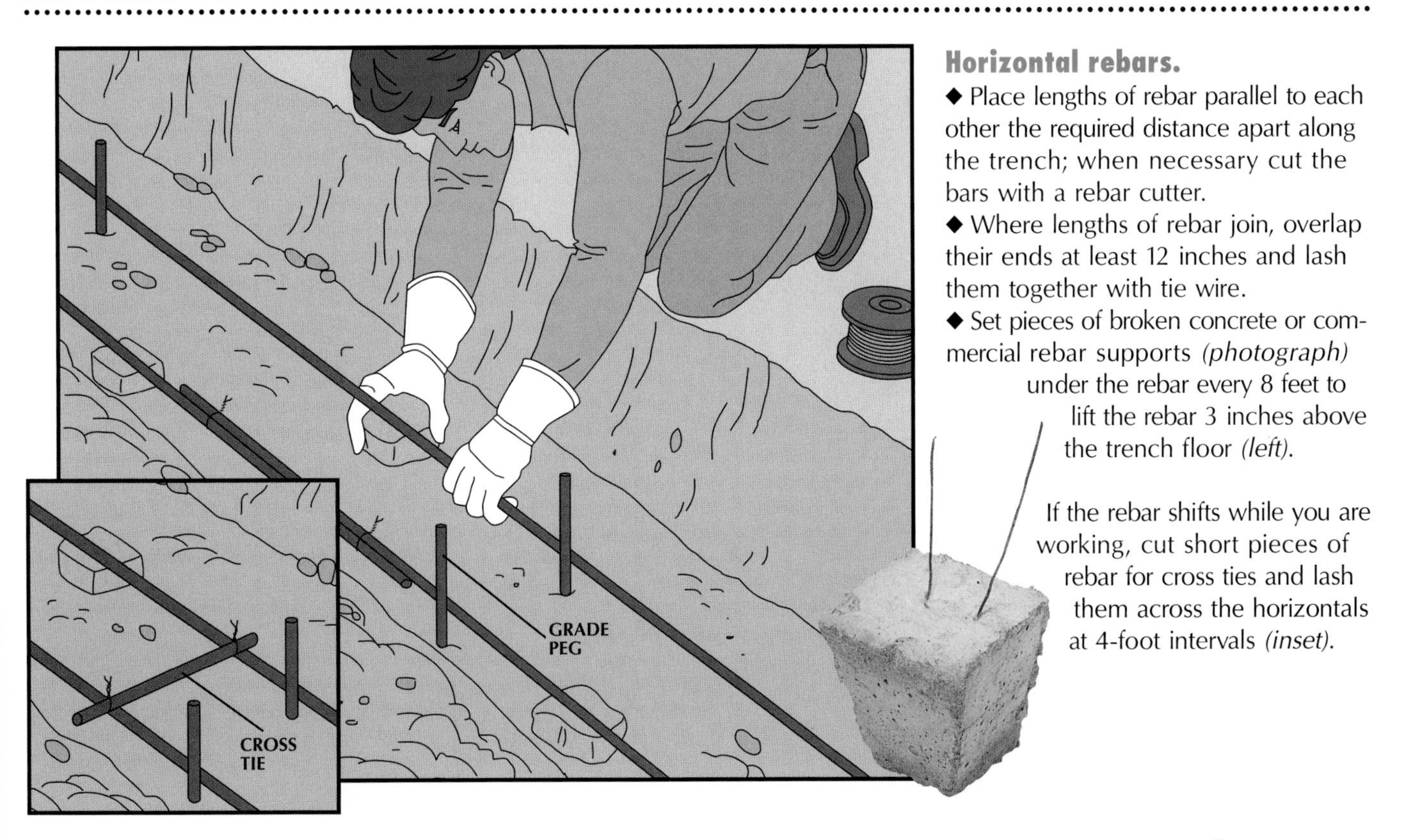


### Horizontal rebars.

◆ Place lengths of rebar parallel to each other the required distance apart along the trench; when necessary cut the bars with a rebar cutter.

◆ Where lengths of rebar join, overlap their ends at least 12 inches and lash them together with tie wire.

◆ Set pieces of broken concrete or commercial rebar supports *(photograph)* under the rebar every 8 feet to lift the rebar 3 inches above the trench floor *(left)*.

If the rebar shifts while you are working, cut short pieces of rebar for cross ties and lash them across the horizontals at 4-foot intervals *(inset)*.

**Vertical rebars for extra strength.** Where codes require the building wall to be reinforced, you will need to cast vertical lengths of rebar into the footing. Long enough for three courses of concrete block to be laid over them when the wall is built, they must extend 20 inches up so that a second piece can be tied to the first.

◆ Cut two short pieces of rebar for cross ties and lash the ties across the horizontal rebar 5 inches apart, where the vertical rebar will sit.

◆ With a rebar bender *(box, below)*, bend a bar to 90 degrees at a point 6 inches from one end.

◆ Cut off the opposite end so the rebar is long enough to extend 44 inches above the top of the footing.

◆ Lash the vertical rebar to the cross ties as tightly as possible with tie wire *(right)*; it can be straightened when the concrete is poured.

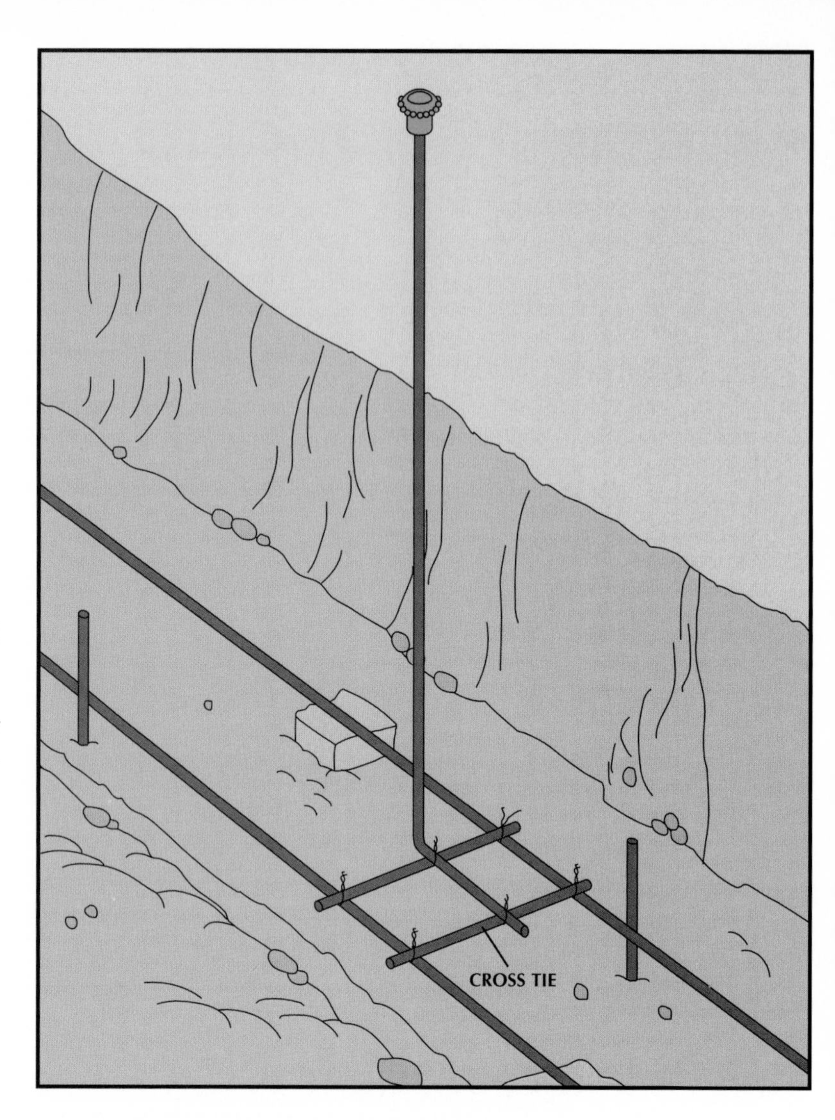


## BENDING REBAR

A heavy-duty commercial rebar bender makes it simple to shape bars to any angle. Line up the bender arm with the roller and slide the rebar into place across the two. Position the handle in the hole most convenient for the angle required, then push it down until the rebar is bent to the correct angle *(below)*. The model shown has a built-in rebar cutter in the center of the roller. To cut rebar, slide the rod in the cutting hole and press the handle down.

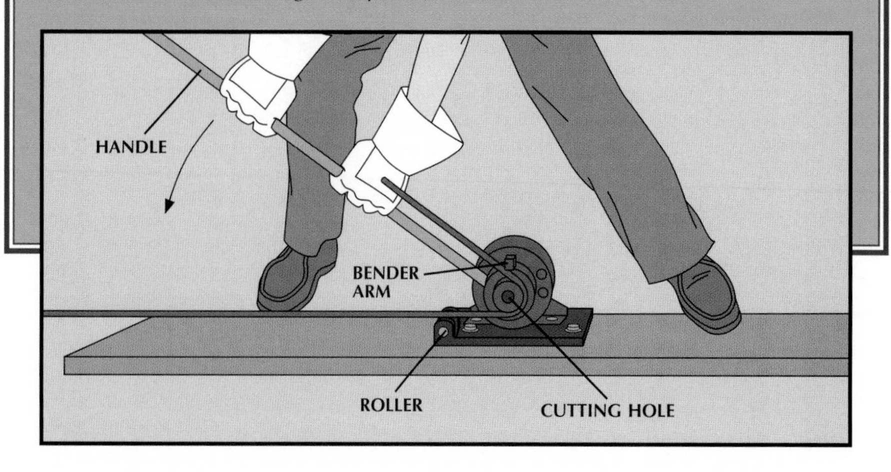

**Reinforcing a stepped footing.**

◆ Measure the length and height of each excavated step and, with a rebar bender, shape the bar to match the contours of the steps.

◆ Lay the lengths of bent rebar the required distance apart.

◆ Prop the horizontal sections with pieces of concrete or rebar supports so that they are 3 inches off the excavation floor, and adjust the pieces so the vertical sections are 3 inches away from the vertical faces of the steps *(right)*.

◆ Join pieces of rebar only on the horizontal run, overlapping the pieces by 12 inches and lashing them together with tie wire.

◆ Install a baffle at each step *(below)*.

# MAKING STEPS WITHOUT FORM BOARDS

**1. Marking the locations.**

◆ Measure out 2 feet in front of the excavated step, and mark a chalk line 12 inches long at a right angle to the trench *(left)*. Repeat the process to mark the opposite side and all the other steps in the same way.

◆ Drive 20-inch 1-by-4 support stakes diagonally into the sides of the trench at every mark, sinking the stakes at least 12 inches into the earth.

◆ Cut $\frac{1}{2}$-inch plywood baffles 3 inches longer than the width of the trench at each set of marks, and as wide as the step will be high.

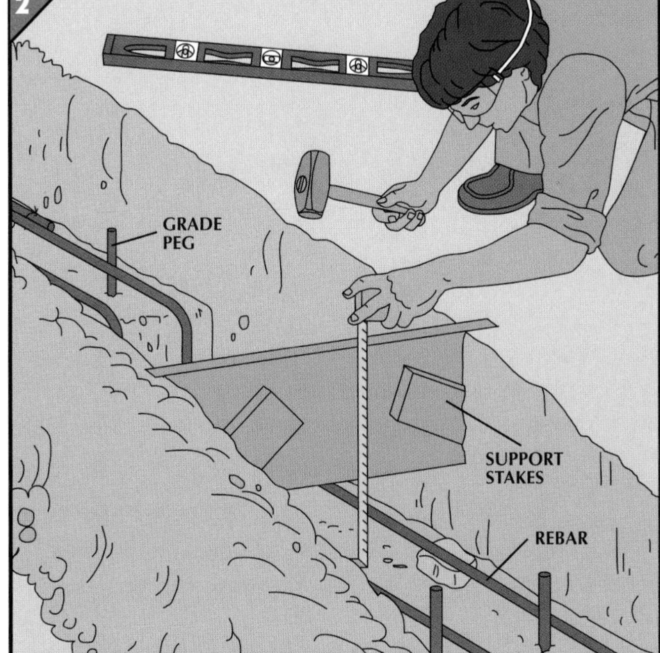


**2. Leveling the baffles.**

◆ On the uphill side of the support stakes, tap the baffle with a maul to drive it into the sides of the trench, stopping when the bottom edge of the baffle is at the height of the top of the planned footing *(right)*.

◆ Hold a carpenter's level against the face of each baffle to check it for plumb and against the top edge to check it for level, and make any necessary adjustments to plumb and level the board.

◆ Drive in a baffle at each set of support stakes, then, bracing the back of the board with a maul, fasten the stakes to the boards with 2-inch double-headed nails.

## POURING THE CONCRETE

### A level footing.

◆ While a helper pours concrete into the trench, spread it evenly between the trench walls or the wooden forms with a square shovel, working it underneath the rebar and chopping it lightly to get rid of any air pockets *(right)*.
◆ Continue pouring until the concrete just reaches the top of the grade pegs or forms.
◆ If there are forms, tap the outside of the boards with a hammer to help settle the concrete.

### A stepped footing.

◆ Have a helper fill the lowest level of the trench, pausing partway for you to work the concrete around the rebar and beneath the first baffle with a square shovel.
◆ Continue pouring until the concrete reaches the tops of the grade pegs or form boards, then push the concrete under the baffle and chop it lightly to get rid of any air pockets.
◆ Pour the concrete for the second step in the same way as for the first. Push the concrete under the second baffle and pull it against the first baffle until it is level with the baffle top and with any grade pegs *(left)*.
◆ Continue pouring and leveling the concrete into each step up the slope using the same technique until the entire footing is poured.

## FINISHING THE FOOTING

### Smoothing the surface.

◆ Smooth the concrete with a float until the tops of the grade pegs just show *(right)*; or, if you are working with forms, first level the concrete with the top of the boards with a 2-by-4 long enough to span the form's top edges, then smooth the concrete with a float.
◆ If the code specifies that the grade pegs need to be removed, pull them out with pliers, and fill the resulting holes by plunging a hand gardening spade halfway into the concrete at each hole; then repair the surface with the float.
◆ Let the footing set 72 hours, covered loosely with plastic sheeting.

# A Slab for a Combined Footing and Floor

Most small single-story buildings, such as garages and garden sheds, can be built on a concrete slab that combines the footing with the floor. Called a turned-down or thickened-edge slab, this dual-purpose construction can incorporate insulation and, if permitted by code, floor drains *(pages 52-53)*. Grade beams can also be built in as reinforcement, allowing you to carry the slab over marshy ground or unstable soil. Ideally, it is best not to build over shallow tree roots; if there are roots that you do not wish to cut, you can place grade beams over them *(page 56)*.

**Planning the Job:** Before you begin the job, verify that local soil conditions and building codes permit this type of construction, and find out the specific requirements for the thickness of the slab and the depth of the perimeter trench.

Consult the building code to determine the diameter of the drainpipe and the depth and slope of the trench leading to the main drain. The code will also give specifications for the size of wire reinforcing mesh for the slab and rebar for the footing—and whether they are necessary at all—and the thickness of the insulation panels required if you plan to heat the structure.

A building-supplies dealer can estimate how much rebar, tie wire, and woven wire mesh you need for the job. Calculate the amount of gravel and concrete for both the footing and slab with the formula for concrete on page 44, then get the mixture recommended by a dealer delivered in a transit-mix truck.

**Preparing for the Pour:** Site layout and rough-grading are done in the same way as for a standard slab *(pages 34-43)*. Unless the soil has a high clay content and is very compact, the slab will need wood forms *(pages 52-53)*. Have several helpers on hand to quickly pour and shovel the concrete into the form, then into the footing trench. If the concrete is being poured directly from the truck, begin working at the farthest corner of the form.

**CAUTION** *Before excavating trenches, find the locations of underground obstacles such as dry wells, septic tanks, and cesspools, and electric, water, and sewer lines.*

**TOOLS**

- Shovel
- Maul
- Circular saw
- Hammer
- Handsaw
- Tamper
- Rebar cutter
- Rebar bender
- 2 x 4 screed
- Utility knife
- Bolt cutters
- Wheelbarrow
- Square shovel
- Bull float
- Float
- Steel trowel

**MATERIALS**

- 1 x 4s, 2 x 2s
- Form boards (2")
- Exterior-grade plywood ($\frac{1}{2}$")
- Common nails (1$\frac{1}{2}$")
- Double-headed nails (3")
- Drainpipe, elbow, and extension
- Polyethylene sheeting (6-mil)
- Rebar and supports
- Tie wire
- Gravel
- Insulation panels
- Wire mesh
- Concrete
- Anchor bolts
- Rebar stirrups

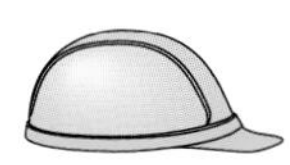

**SAFETY TIPS**

*Wear goggles and gloves when pouring and spreading concrete, and rubber boots to walk in wet concrete. Add a hard hat and work boots when operating a power trencher. Put on work gloves to cut wire mesh.*

## POWER TRENCHER

Available at tool rental centers, a power trencher provides an alternative to trenching by hand and enables you to dig with relative ease. It has a digging chain on a boom that can be raised and lowered according to the depth required for the trench, and cutters can be added to the boom or removed from it, depending on the terrain in which you are working; add cutters for easy removal in finer dirt, remove cutters in rocky ground. Set the trencher to single-wheel drive when moving it; two-wheel drive provides traction when digging. For safety, an automatic shut-off mechanism kicks in when the operator lets go of the hand grips.

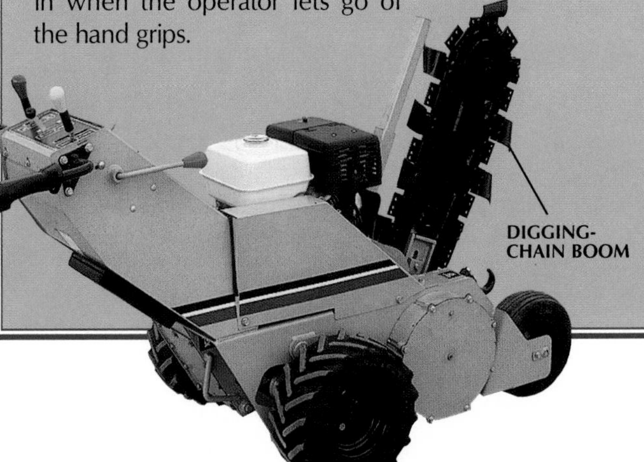

# CASTING THE SLAB

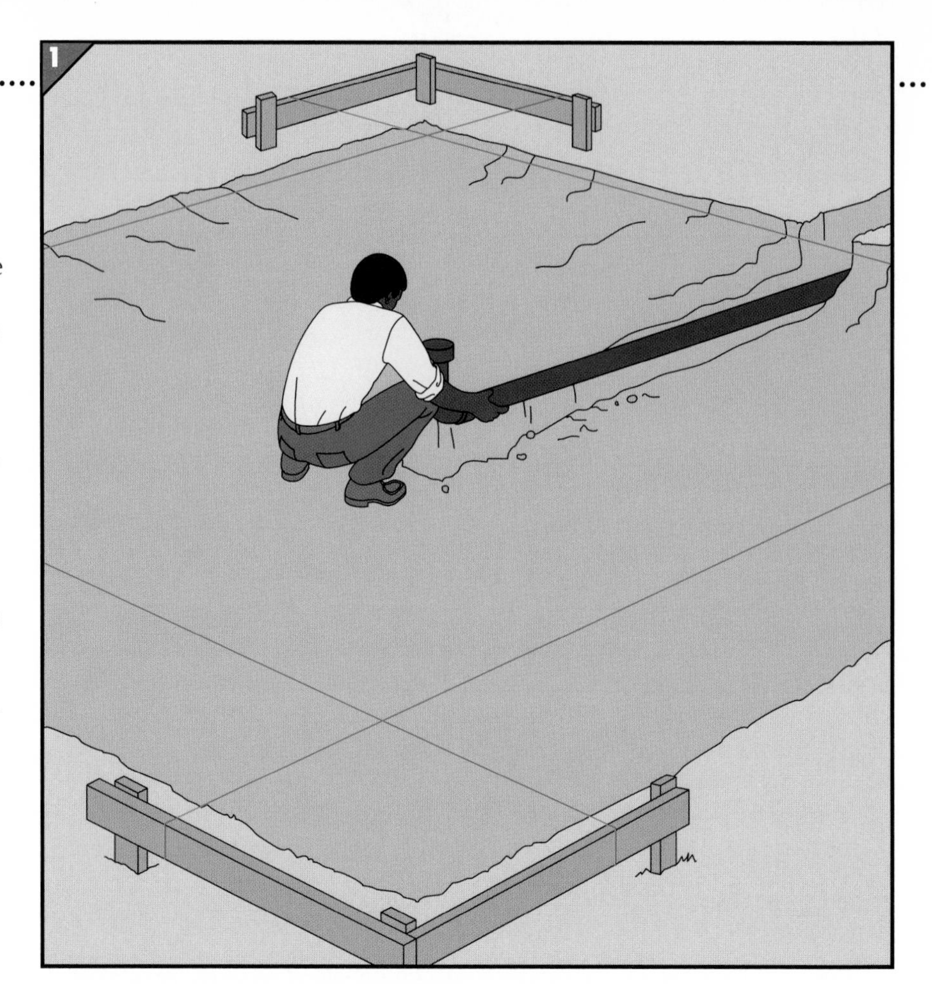


**1. Preparing the site.**

◆ Excavate the slab area according to local building codes.

◆ Drive a stake in the center of the excavation to mark the location of the drain opening.

◆ Dig a trench of the required depth and slope from this point to the nearest main drain or drainage ditch, then remove the marking stake.

◆ Attach an elbow fitting to a length of drainpipe and set the assembly in the trench *(right)*.

◆ Erect outside form boards *(page 45, Step 1)*, aligning their inner faces with the building lines and their top edges with the planned height of the slab.

◆ At 3-foot intervals brace the forms, as for a standard footing, with outrigger stakes *(page 46, Step 2)*.

**2. Adjusting the drainpipe.**

◆ Between two opposite form boards, stretch a string that passes over the drainpipe opening, securing the string with nails at the top of each form.

◆ Measure the distance from the form board to the drain location, then calculate how far down from the string the top of the drain must be to allow for a downward pitch of $\frac{1}{4}$ inch per foot from the form to the drain.

◆ To determine the length of vertical pipe extension needed to bring the drain up to the correct height, measure from the top of the elbow to the string, then subtract the figure calculated above.

◆ Add the pipe extension to the elbow; if necessary dig out or fill in the trench to bring the top of the drain to the correct height *(left)*.

◆ Fill in the trench to the level of the excavation, and tamp it firmly.

◆ Tape plastic over the drain opening to keep it free of concrete during the pouring.

◆ Brace the rim of the drain with two $\frac{1}{2}$-inch rebar spikes, driven into the ground on opposite sides of the drain, just deep enough for the rim to rest on.

◆ Remove the string.

**3. Digging and reinforcing.**

◆ Using the inner face of the form boards as a guide, dig a trench to the depth specified by the building code around the perimeter of the slab area; angle the inner face of the trench so that the trench is 12 inches wide at the top and 7 inches at the bottom.

◆ If rebar is required by code, lay two pieces of the recommended size in the trench 6 inches apart and supported on pieces of concrete or on rebar supports that raise the bars 3 inches above the bottom of the trench. Place the supports at 4-foot intervals and bend the rebar at the corners. Overlap the pieces by 10 inches.

◆ Fasten the horizontal sections of rebar together with tie wire *(right)*.

◆ Cover the slab area, but not the trench, with a 3-inch layer of gravel.

◆ Level the gravel with a 2-by-4 screed.

**4. Insulating the slab.**

◆ Lay the insulation panels along the inner face of the trench, cutting them so that their tops lie level with the gravel surface *(left)*.

◆ Anchor the panels at the bottom of the trench, if necessary, with dirt.

◆ Lay additional insulation along the edge of the slab area, extending inward about 2 feet.

◆ Spread wire reinforcing mesh over the entire slab area and trench, positioning pieces of concrete under it at 3-foot intervals at a height that will place the mesh at the center of the slab. (For a 4-inch slab, set the mesh 2 inches above the gravel.) Position the mesh so that it fits around the sides of the drain, and ends about 2 inches in from the form boards.

◆ Cut diagonally into the mesh at the corners with bolt cutters and bend the corners down, following the shape of the slab corner.

◆ Lash the mesh to the rebar with tie wire *(inset)*.

**5. Filling the forms with concrete.**

◆ Pour the concrete and spread it with a square shovel *(page 50)*, working in wedge-shaped sections; as soon as one section is filled, level the concrete with a 2-by-4 screed.

◆ To slope the concrete surface down to the drain, have a helper hold one end of the screed in position on the drain and rest the other end atop the form boards. Grasp the end of the screed and pull it in an arc with small back and forth movements *(above)* to smooth the surface and establish its downward slope.

◆ Fill in any depressions with wet concrete, then screed them.

TRICKS OF THE TRADE

## Screeding Solo

Leveling concrete is more often—and more easily—done by two people, but if you are working alone, you can make a screed for the job. Cut a 1-by-4 of a lightweight wood such as cedar to a 10-foot length. Cut two 2-foot-long 1-by-2 handles with a 45-degree angle at one end, and screw the handles to the long board, spaced a comfortable distance apart. The angled handles will help reduce the strain of leaning forward while working.

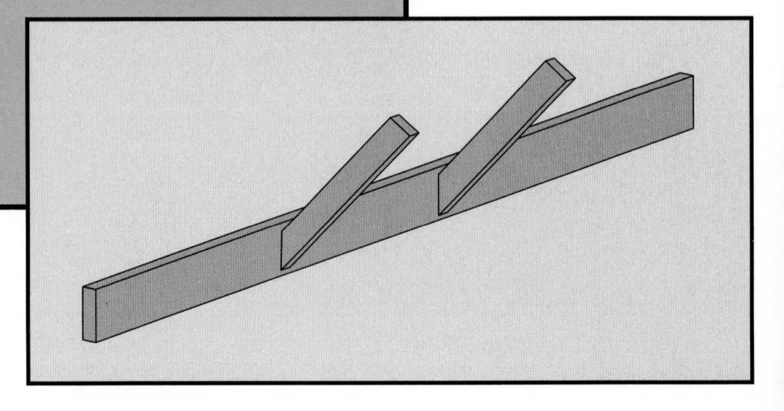

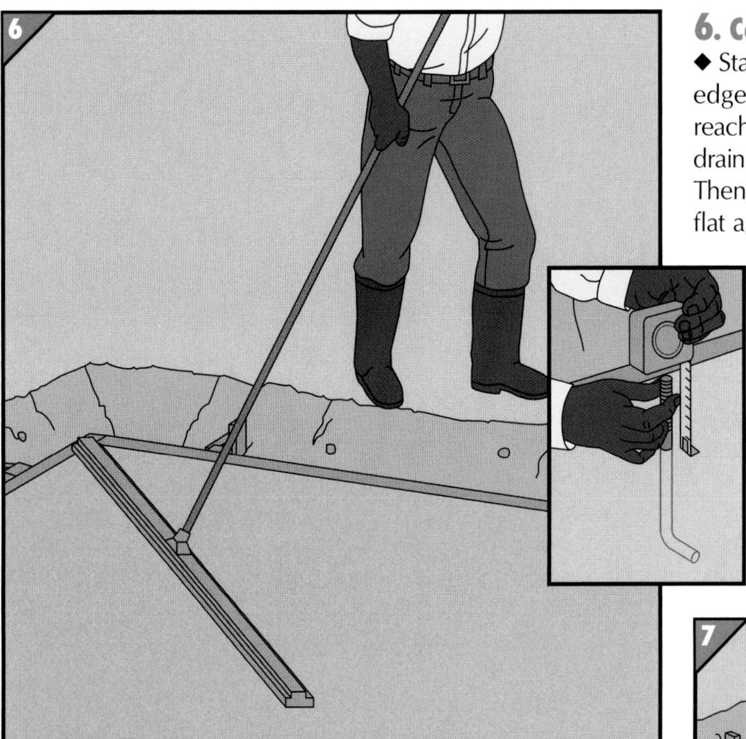

## 6. Compacting the surface.

◆ Standing just outside the slab, position a bull float near the edge of the slab, then push it in an arc as far as you can reach, with the length of the blade always pointing toward the drain and with the front edge of the blade tilted upward *(left)*. Then, draw the float back around the arc, keeping the blade flat against the concrete surface.

◆ Continue to work in wedge-shaped sections in the same way, gradually moving from the edge of the slab to the center, until you have floated the entire slab.

◆ Set anchor bolts along the edges of the slab, spacing them according to code; position them in from the form boards at a distance that is half the width of your planned sill plate—the heavy board that underpins a wood-frame structure—and so that the top of the bolt protrudes by $2\frac{1}{2}$ to 3 inches *(inset)*.

◆ Allow water on the surface of the slab to evaporate before you continue.

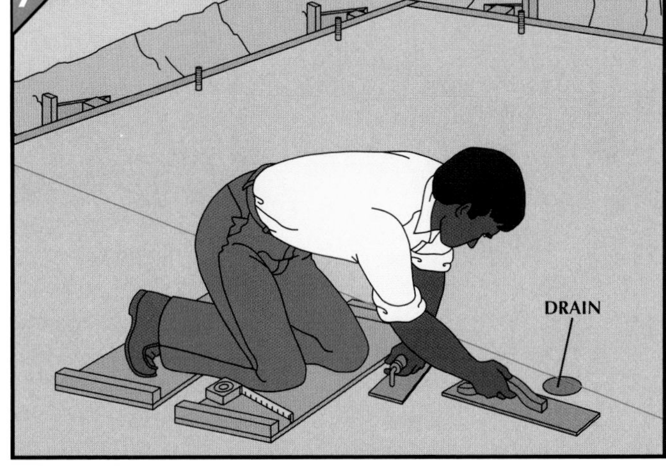


## 7. Finishing the slab.

◆ Stretch a string across the slab so that it passes over the drain *(page 52, Step 2)*.

◆ Construct a pair of kneeboards so you can move around on the slab by nailing 2-by-2 handles to each end of rectangles of $\frac{1}{2}$-inch exterior-grade plywood.

◆ Position yourself near the drain with your knees on one knee board and your feet on the other. Smooth the concrete surface into a gradual slope by sweeping the surface in overlapping arcs, first with a float, then with a steel trowel *(right)*.

◆ Working out toward the edge of the slab, reposition yourself as necessary to smooth one quadrant of the slab. Every few feet, check the slope by measuring from the string to the surface to ensure that the surface drops about $\frac{1}{4}$ inch every foot. Fill in any depressions with wet concrete.

◆ Repeat the smoothing process for the remaining quadrants.

◆ Wet the smoothed slab with a fine spray of water and cover it with overlapping sheets of 6-mil polyethylene.

◆ Let the concrete cure for three days, then remove the plastic and the forms.

### A POWER TROWELER FOR QUICK FINISHING

The task of floating and finishing concrete surfaces can be simplified with a power troweler *(photograph)*. To use the machine, start at one corner of the surface and move the troweler from side to side as you walk backward, stepping off the concrete at the corner opposite your starting point. Never let the troweler stand too long in one spot, and to avoid creating valleys in the slab surface, alternate the direction of each complete pass over the slab, running back and forth first over its length, then over its width. The angle of the blades, as well as the speed at which they turn, is easily adjusted to suit the job. Start troweling with the blades flat and rotating at a slow speed when the concrete is wet, gradually increasing speed as the surface hardens. Increase the pitch angle of the blades slightly to produce a smoother surface.

Avoid using a power troweler on air-entrained concrete of 3 percent or more; air pockets can become trapped just below the surface, reducing the concrete's durability.

# REINFORCING WITH GRADE BEAMS

### A slab for shifting soil.

◆ After reinforcing the edges of the slab area *(page 53, Step 3)*, dig trenches for grade beams at the intervals, width, and depth specified by code, laying them out parallel to two sides of the footing trench.
◆ Lay horizontal rebar of the recommended diameter in the grade-beam trenches as you would for a footing *(page 47)*, bending it outward at either end so it overlaps the footing rebar by 10 inches, and tying it to the footing rebar with wire.
◆ With tie wire, attach a rebar stirrup to each piece of rebar at every support and two at each end of the rebar, attaching them to the footing rebar.
◆ Tie a second layer of rebar in the stirrups, above the first *(inset)*.
◆ Spread a 3-inch layer of gravel in the slab area, but not in the trench, and screed it with a 2-by-4.
◆ Place insulation and wire mesh in the excavation *(page 53, Step 4)*, then tie the grade-beam rebar to the mesh at 2-foot intervals.
◆ Pour and finish the slab *(pages 54-55, Steps 5-7)*.

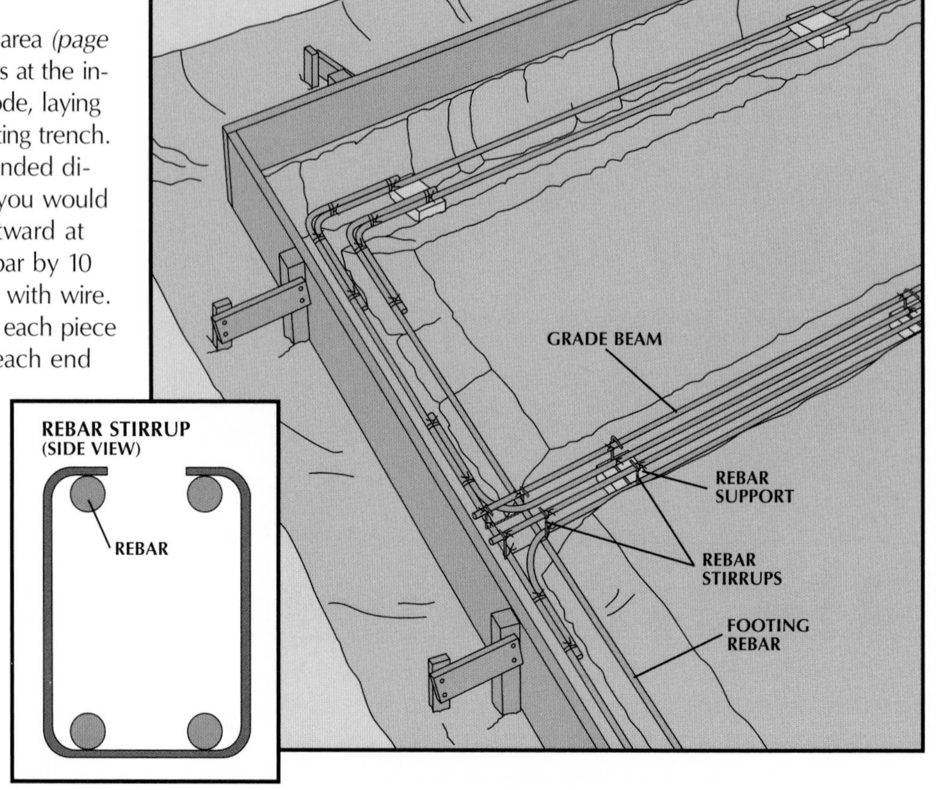


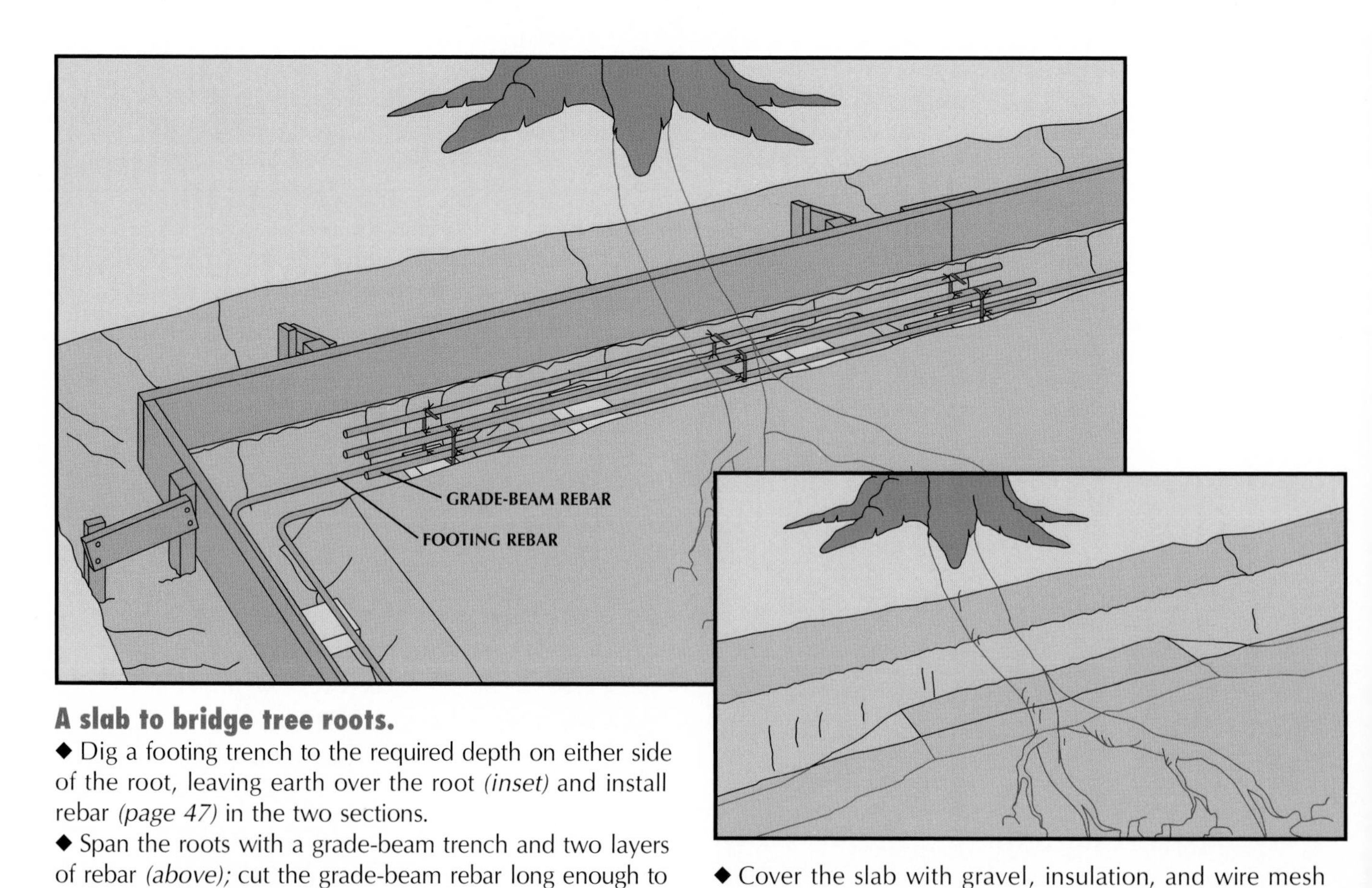


### A slab to bridge tree roots.

◆ Dig a footing trench to the required depth on either side of the root, leaving earth over the root *(inset)* and install rebar *(page 47)* in the two sections.
◆ Span the roots with a grade-beam trench and two layers of rebar *(above)*; cut the grade-beam rebar long enough to overlap the footing rebar by 15 inches, and join the three layers of rebar with stirrups as for an ordinary grade beam.
◆ Cover the slab with gravel, insulation, and wire mesh as for a regular slab, then tie the rebar in both the grade-beam trench and the footing trench to the wire mesh.

# Shaping Concrete into Decorative Blocks

Individual concrete blocks can be made in a variety of shapes for decorative structures such as low walls, garden walkways, and borders for flower beds. Color, pattern, and texture can be added for further visual appeal. These ornamental blocks are poured directly into cavities scooped out of the earth *(below)* or cast in removable wooden forms *(page 58)*. The concrete is poured, packed, and leveled in the same way as for a concrete footing *(page 50)*, with some modifications for the type of decorative surface desired. You can mix the concrete yourself in a wheelbarrow or a mortar pan, adding water to premixed concrete or to a basic mix of one part cement, two parts sand, and three parts coarse aggregate.

**Textures and Patterns:** Forms can be lined with materials such as ridged rubber matting *(page 59)*, although materials with an undercut pattern may interfere with removing the block from the mold. In climates where alternate freezing and thawing occur, avoid surface patterns that will collect rainwater.

Attractive aggregates, such as polished pebbles and marble chips, added to the concrete mix or sprinkled on top of plain concrete, will create striking textures *(page 60)*. Aggregate smaller than $\frac{1}{4}$ inch in diameter is more versatile for creating texture and pattern. Other decorative looks can be produced with salt crystals *(page 60)*, or by roughening the surface to resemble travertine *(page 61)*.

**Color:** Although precolored cement for mixing with concrete is available, it is cheaper to add mineral-oxide pigments yourself—and the color choice is wider. To achieve the clearest colors, work with white cement, white sand, and white aggregate. Mix colored concrete in an automatic mixer to avoid any blotchiness, adding half the usual water until the color is uniform. For a pastel look, the proportions are 1 or 2 pounds of pigment to every 100 pounds of cement; for deep shades, a good ratio is 7 pounds of pigment to every 100 pounds of cement.

Always measure the ingredients by weight rather than volume, and never add more than 10 percent pigment, or it may weaken the concrete. You can make several small test batches; let these miniature blocks cure for about a week, as their color will change slightly in curing. Keep a record of ingredients, so that you can duplicate colors.

The most economical way to apply color is with a mixture called dry-shake—a combination of pigment, cement, and sand that is sprinkled on top of the still-damp concrete block and floated in.

**TOOLS**

- Shovel
- Spade
- Float
- Handsaw
- Hammer
- Screwdriver
- Mason's trowel
- Stiff brush
- Mason's utility brush
- Nail set

**MATERIALS**

- 1 x 2s, 2 x 3s, 2 x 4s
- Plywood ($\frac{3}{4}$")
- Quarter-round molding
- Common nails (2", $3\frac{1}{2}$")
- Finishing nails
- Hinge
- Hook-and-eye
- Concrete mix
- Form oil
- Rubber matting
- Gravel
- Polyethylene sheeting
- Wood putty
- Pebbles or stones
- Rock salt
- Mortar mix: Portland cement, sand

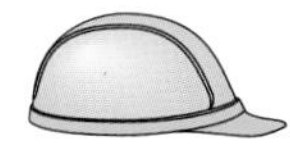

**SAFETY TIPS**

*When mixing, pouring, and leveling concrete, wear gloves and protect your eyes with goggles.*

## EARTH AND WOOD FORMS

### Earth forms for stepping-stones.

◆ Dig holes 3 inches deep in the desired shapes of the blocks.

◆ Cut the edges of the holes with a spade or a garden trowel so that the perimeters are clean and as near to vertical as possible.

◆ Mix concrete and fill one of the holes, then tamp and smooth it with a wood float, leaving a slightly rough but uniform texture; if the stepping-stones are in a lawn that will need to be mowed, make sure their top surface is level with the surrounding earth *(right)*.

◆ Pour and finish each of the remaining stepping-stones in this way.

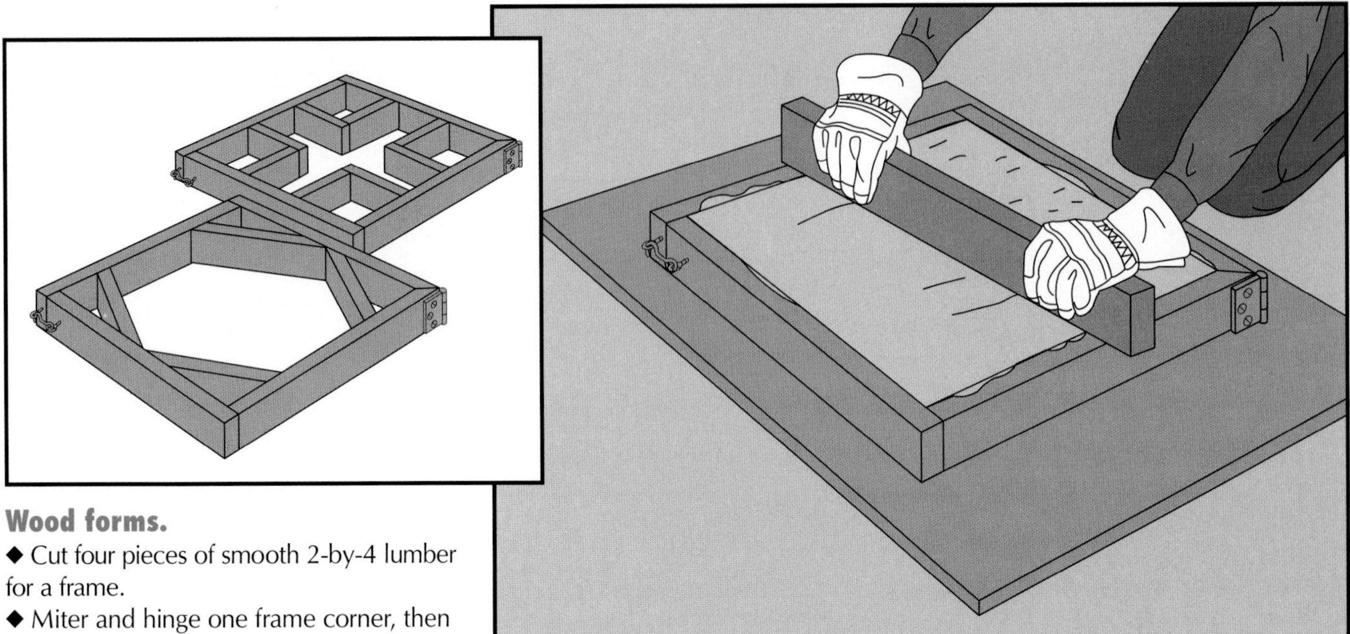

### Wood forms.

◆ Cut four pieces of smooth 2-by-4 lumber for a frame.
◆ Miter and hinge one frame corner, then butt-join the two adjacent corners with $3\frac{1}{2}$-inch common nails; close the fourth with a hook-and-eye.
◆ Set the frame on $\frac{3}{4}$-inch plywood, and coat the plywood and frame lightly with form oil.
◆ Fill the form, then tamp and level the concrete with a 2-by-4 *(above).*
◆ Run a mason's trowel between the concrete and the form to compact the edges, then smooth the surface with the trowel.
◆ Leaving the block in place on the plywood, remove the frame after 10 minutes, wash it with water, and repeat the process to make more blocks.
◆ Let the blocks cure for 24 hours before moving them, then for another six days under plastic.

To make interlocking blocks, add removable inserts to a square frame: L-shaped pieces for a cruciform block; mitered corner pieces for a hexagon *(inset).* For the blocks to fit together, all sides of the form must be the same length.

### Ganged forms.

◆ Construct a rectangular frame of 2-by-3s with $3\frac{1}{2}$-inch common nails, fastening the end pieces 1 inch in from each end; nail 1-by-2s to the end pieces with 2-inch nails to make handles.
◆ Install 2-by-3 dividers inside the frame to define compartments of the desired shapes.
◆ Prepare the area where the blocks will be placed, digging a trench if desired.
◆ Lightly oil the form, and cast the concrete as you would for an individual form *(above),* but level all the blocks at once by pulling a 2-by-4 over their surface.
◆ After at least 15 minutes, remove the form by lifting it straight up off the blocks *(left).*
◆ Reposition the form and repeat the procedure.
◆ Let the blocks cure for 24 hours before moving them, then for another six days under plastic.

# CREATING PATTERNS WITH FORM LINERS

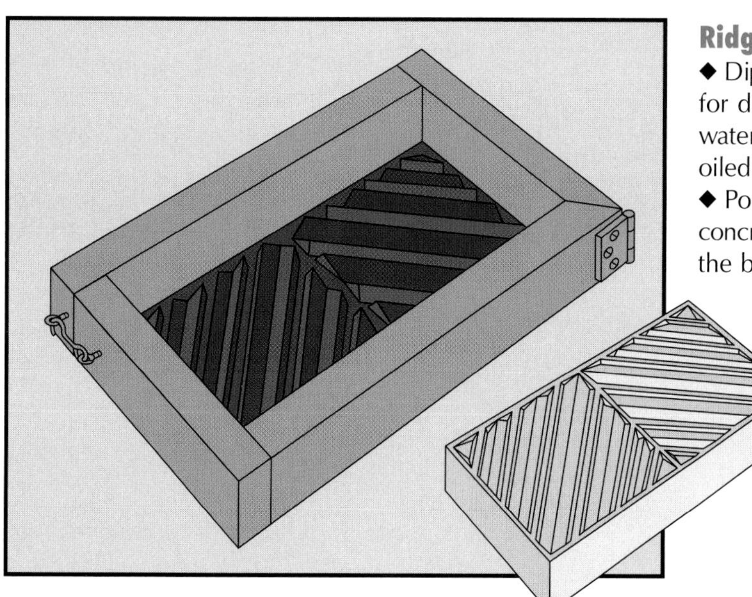

**Ridging with matting.**

◆ Dip rubber matting—normally sold for doormats and stair treads—in water and place it beneath a lightly oiled form *(left)*.

◆ Pour in a relatively wet mix of concrete, then tamp, level, and cure the blocks *(opposite)*.

**Embossing with gravel.**

◆ Spread gravel of fairly uniform size evenly over an area that matches the internal dimensions of the form, leaving spaces between the individual stones.

◆ Lay a sheet of household plastic wrap or polyethylene sheeting loosely over the gravel so that the weight of the concrete will force the film into the spaces.

◆ Position a lightly oiled form over the plastic *(left)*, pour in a relatively wet mix of concrete, then finish the concrete *(opposite)*.

**Shaping with molding.**

◆ Cut quarter-round molding the length of each of the sides of the form, and miter the ends so that they will fit snugly together at the corners.

◆ With finishing nails, attach the molding along an inside edge of the frame with the quarter-round facing inward; set the nails in with a nail set and fill in the holes with wood putty.

◆ Lightly oil the form and set it on a piece of $\frac{3}{4}$-inch plywood, also lightly oiled *(left)*.

◆ Pour in a relatively wet mix of concrete, then finish it *(opposite)*.

**Exposed aggregate.**
◆ Mix concrete, using the desired decorative aggregate, then pour it into an oiled form.
◆ Tamp and level the concrete with a 2-by-4 as you would for a plain block in a wood form *(page 58)* and let it set until the aggregate is firmly anchored but the concrete is still soluble—about 1 hour.
◆ With a stiff brush, test a corner of the block to be sure that brushing will not dislodge the aggregate.
◆ With a garden hose and the brush, simultaneously flush and scrub the top of the block until the top of the aggregate is exposed *(left)*.

**Pebbled paving.**
◆ Trim a 2-by-4 to the width of the form, then cut a $\frac{1}{4}$-inch-deep notch at each end so it will fit over the edges of the form.
◆ Fill the oiled form with concrete to within $\frac{1}{4}$ inch of the top, then tamp and level the surface with the 2-by-4.
◆ Wet the pebbles or stones to be added, and distribute them in a single layer over the concrete, pressing them into the concrete with a float until they are buried just below the surface *(right)*.
◆ When the concrete is set but still soluble, flush and brush the surface as for exposed aggregate to reveal the tops of the pebbles or stones.

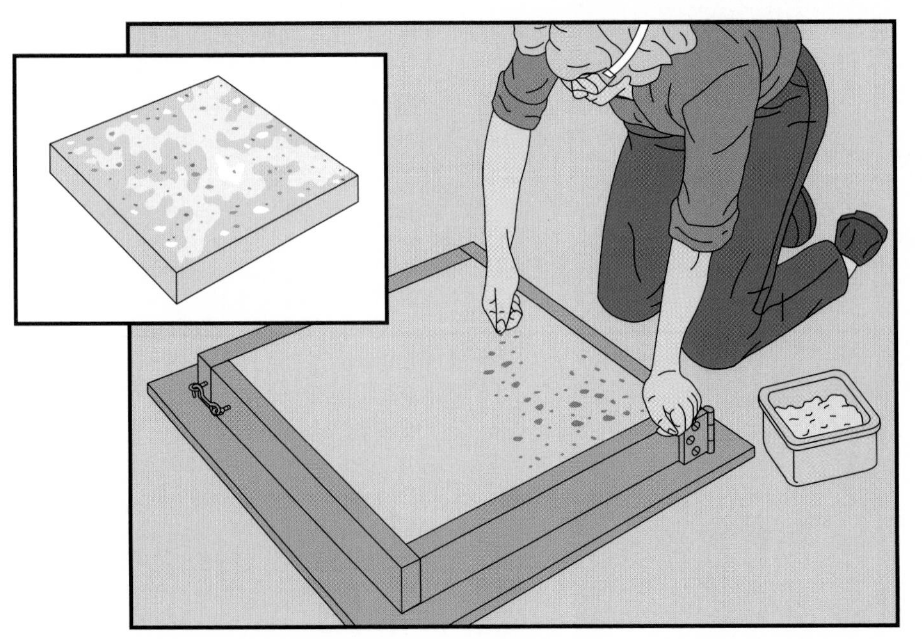

**A salt-pitted surface.**
◆ Scatter large grains of rock salt over the block while it is still damp—just after it has been troweled or floated smooth *(left)*.
◆ Press the salt crystals into the concrete with a float, but do not bury them.
◆ Allow the block to cure, then wash away any undissolved salt with a garden hose.

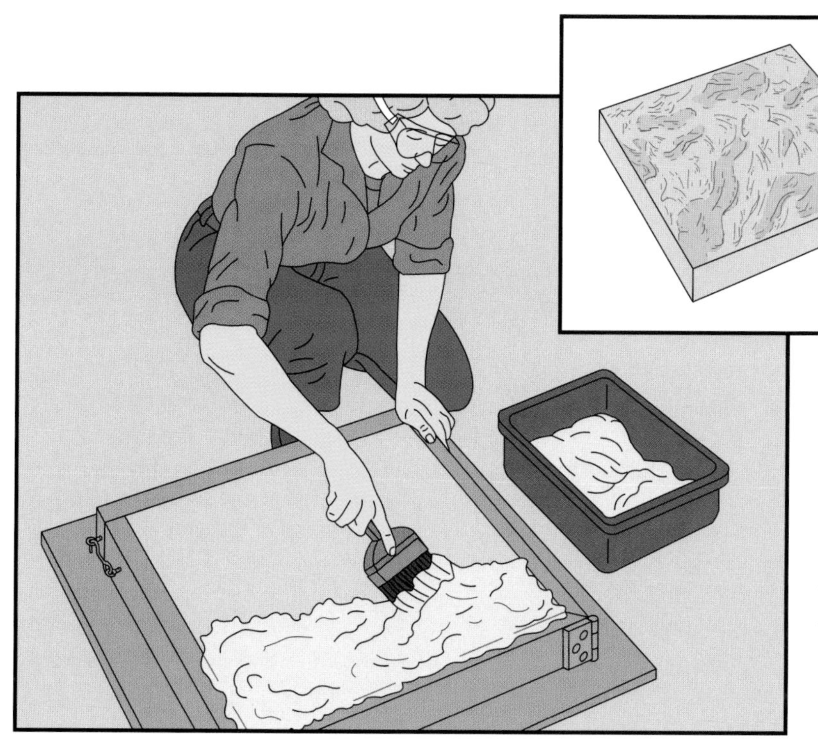

**A travertine finish.**

◆ Tamp and level the surface of the block with a 2-by-4 as you would for a plain block in a wood form *(page 58)*, but smooth it only lightly, so that the concrete retains a rough surface.

◆ With a mason's utility brush, dab on mortar made of two parts Portland cement to one part sand, coating the concrete with an uneven, patchy surface, with ridges up to $\frac{1}{4}$ inch high.

◆ When the water sheen disappears, trowel or float off the tops of the ridges, leaving both mortar and concrete rough in the crevices between.

To heighten the streaked look, tint the mortar topping with mineral-oxide pigment *(page 57)* before you put it on the concrete.

## STAMPING A TEXTURE OR PATTERN

Decorative texture can be added to concrete with a patterned mat or roller. The mat, made of polyurethane, is laid over the surface of freshly floated and finished concrete, and carefully pressed into the concrete with a tamper. These mats are available in a variety of different patterns, such as herringbone, running bond, and cobblestone that give the appearance of brick, tile, or stone *(photograph)*. A quicker way of stamping concrete is with a patterned roller made of rustproof aluminum. The roller is filled with water and carefully guided over the surface of the concrete. The depth of the impression can be reduced or deepened, depending on the effect you want to create.

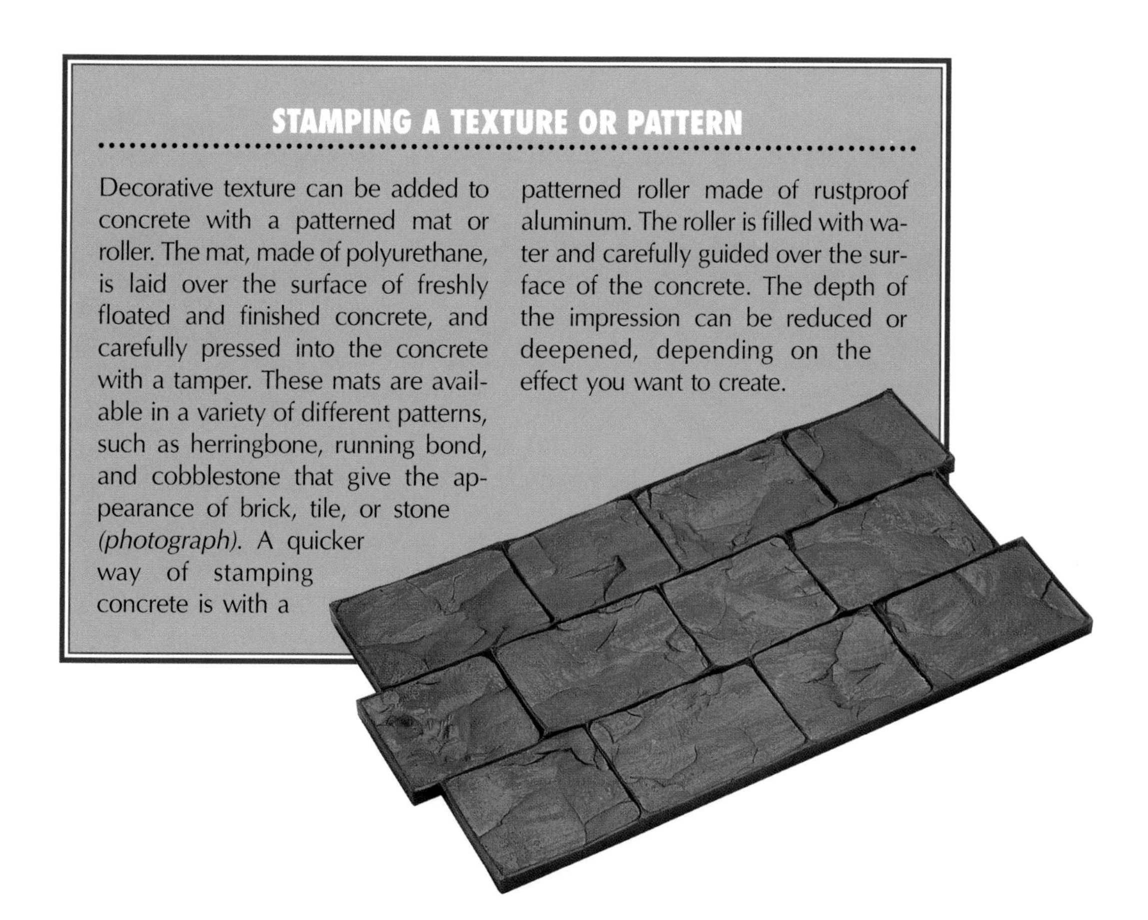

# Making Adobe

**S**tabilized adobe bricks made of earth, water, and Portland cement are a practical building material wherever the right soil is available. Though impervious to rain showers after a few days, they require at least three weeks of dry weather or protection from rain under plastic to harden properly.

**The Right Mixture:** Good soil for adobe consists primarily of sand and some clay; earth with a lot of organic matter will not adhere into bricks. You can easily check dirt for suitability *(below)*. Make a series of test bricks to determine the right mix of cement and water for your soil *(opposite)*. The ratio of cement to soil ranges from about 10 percent for sandy, gravelly soil to 16 percent for soil with some clay.

**Making Bricks:** To combine the materials to make the bricks, rent a power mixer, and mold the bricks in wooden forms *(page 64)*. To estimate how much cement you need for a project, multiply the dimensions of the bricks—4 by 10 by 14 is a good size for a first job—by the number you need, and then by the percentage of cement determined right for your soil in the tests. If you plan to build a structure that is subject to building codes, send test bricks to a commercial lab to be checked for strength, absorption, and moisture content.

**Building with Adobe:** The footing for an adobe structure must extend above ground level. The walls are assembled much like those of standard bricks, and with standard type N mortar, but the mortar joints are thicker—$\frac{3}{4}$ inch.

**TOOLS**

Shovel
Buckets
Awl
Wheelbarrow
Power mortar mixer
Circular saw
Hammer
2 x 4 tamper
Stiff brush

**MATERIALS**

1 x 6s
Cleats
Plywood
Common nails (2")
Steel corner braces
Soil
Portland cement
Polyethylene sheeting
Plastic pipe (3" diameter)

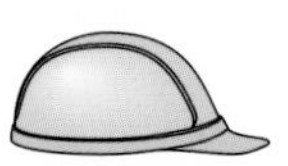

**SAFETY TIPS**

*Gloves protect your hands when working with cement and fresh adobe. Put on goggles when nailing.*

## ANALYZING SOIL CONTENT

**A water test.**

◆ With a shovel, clear away any topsoil, then break and mix the dirt to be tested, removing any large stones.

◆ Place 2 inches of dirt in a quart jar, then fill the jar with water.

◆ Cap the jar and shake it until the contents are thoroughly mixed.

◆ When the soil column has settled and the water has cleared, measure the depth of the smooth top layer of clay and silt, called fines *(right)*. If this layer is $\frac{3}{8}$ inch or less, the soil is suitable. If it is greater than $\frac{3}{8}$ inch, empty the jar and repeat the test, but this time mix one part construction sand with three parts soil. If still more sand is required, making adobe bricks from this soil will be too expensive.

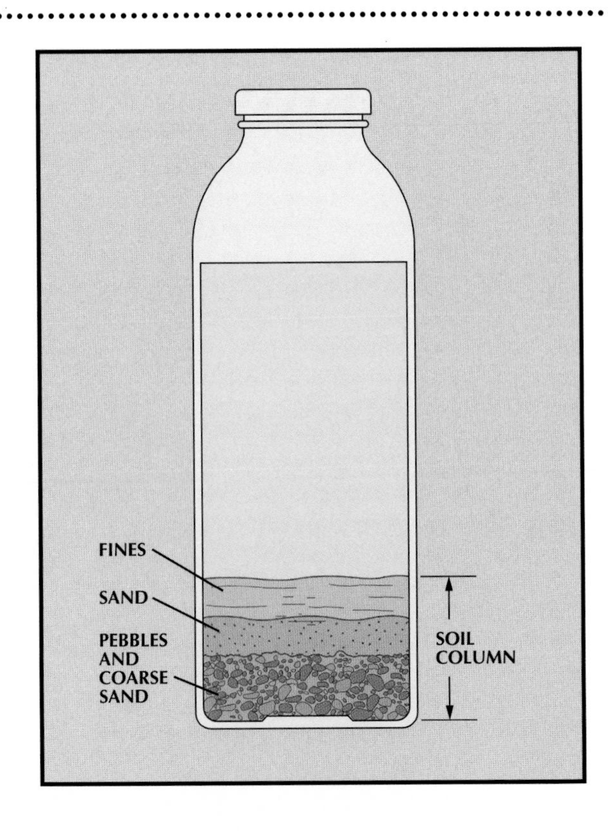

# CREATING TEST SAMPLES

**1. Preparing the samples.**

◆ Collect several buckets of soil from various points.
◆ Measure into a bucket a small amount of soil and cement: If the earth is sandy, use $23\frac{1}{2}$ parts soil and $1\frac{1}{2}$ parts Portland cement. For dirt with some clay, use $22\frac{1}{2}$ parts soil and $2\frac{1}{2}$ parts cement.
◆ Noting the amount as you add it, blend in a little water at a time until the mix is moistened.
◆ Test a handful of the material by squeezing it firmly, then releasing it *(right)*. If water remains on your hand, discard the mixture and make another with less water; if it does not stay together, add more water—keeping note of how much—and repeat the squeeze test.
◆ Form the mixture into a ball and drop it from a height of several feet. If it stays together, discard the sample and mix another with less water; if it shatters, the moisture content is correct.
◆ Prepare eight more samples of this mixture, but adding $\frac{1}{4}$ part more cement to each successive sample and $\frac{1}{4}$ part less soil.
◆ Place one-third of a sample in a container such as a 4-inch length of 3-inch plastic pipe set on a piece of plywood. Tamp the mixture down firmly, then scratch the surface with a fork and add two more layers, tamping and scratching them between layers. Do the same with the seven remaining samples.
◆ Lift off the forms and etch into each cylinder an identification number that represents the cement and water amounts.
◆ Put the samples in plastic bags and let them cure for a week.

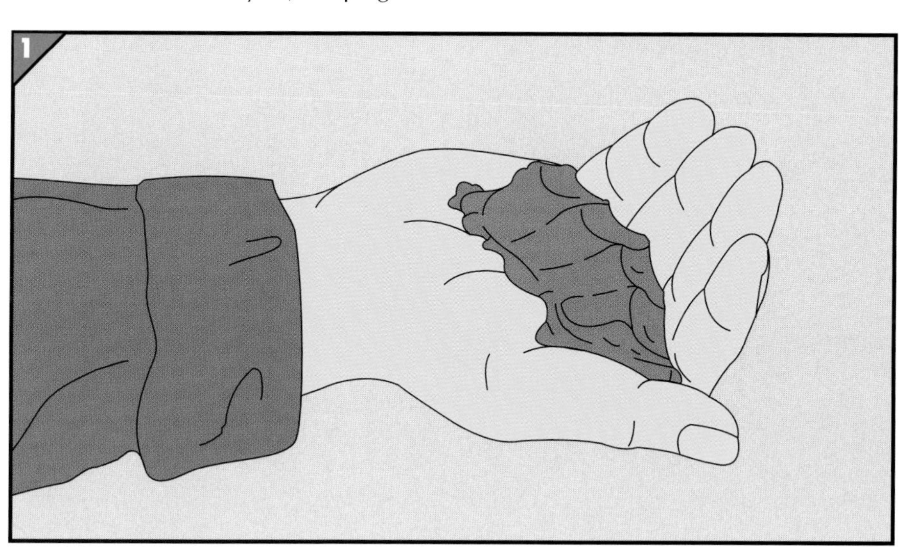

**2. Testing the adobe.**

◆ Soak the dried samples in water for four hours.
◆ Beginning with the one with the least amount of cement, jab it with a blunt awl or ice pick, first lightly then harder. If the point penetrates more than $\frac{1}{8}$ inch but less than $\frac{1}{4}$ inch, keep the sample. Test the remaining cylinders.
◆ From the samples that passed the jab test, knock together two with the lowest amount of cement, gently at first, then harder *(left)*. If you hear a dull thud rather than a clicking sound, at least one piece is not hard enough. Reject the one with the lower cement content, then knock the one the higher amount of cement together with a sample with the next highest amount of cement. Test all the cylinders in this fashion, retaining only the ones that create a clicking sound.
◆ If a cylinder breaks or chips—but does not crumble—during the sound test, repeat the jab test on the broken surface. If it passes the second jab test, it is acceptable.
◆ Let the good pieces dry overnight, then repeat the tests on them.
◆ From the samples that passed all tests, select the one with the lowest amount of cement. Add $\frac{1}{2}$ part cement to the amount used for this sample and subtract $\frac{1}{2}$ part soil to make adobe bricks. (If the bricks will be exposed to foot traffic and weather extremes, add 1 part cement and subtract 1 part soil.)

# FORMING BRICKS

## 1. Mixing the adobe.

◆ The day before you plan to make the bricks, break up any clods of clay in the dirt and mud, and remove any stones. Turn the soil over and mix it, soak it with water, then cover it with polyethylene sheeting.
◆ Add water by the bucketful to a mortar mixer, noting the number of buckets, until there are a few inches in the hopper.
◆ Keeping track of the amount, add soil to the hopper until it is almost at maximum drum capacity *(right)*.
◆ Add the rest of the quantity of water required for the amount of soil being mixed.
◆ Run the mixer until all lumps disappear, then slowly add the required amount of Portland cement while the paddles are turning.
◆ Keep mixing until the material is uniform in color and texture, then turn off the mixer and double-check the moisture content with the squeeze test *(page 63, Step 1)*. Make any necessary adjustments, then tilt the hopper with the dump handle, lock it in position, and shovel the mud into a wheelbarrow.

CAUTION *Do not operate an electric mixer in damp conditions, and cover it when not in use. Never reach into the mixer with your hands or a tool while it is operating. To fuel a gas-powered engine, first turn off the engine and allow it to cool down.*

## 2. Molding the bricks.

◆ Rip smooth 1-by-6s to $4\frac{1}{4}$ inches—the bricks will shrink to about 4 inches when they dry—and construct a frame divided into compartments of the desired size; fasten the boards with 2-inch common nails and reinforce them with steel corner braces. Add a wooden cleat to each end to serve as a handle.
◆ Wet the form thoroughly and place it on plywood or building paper laid on level ground.
◆ Tip the wheelbarrow and shovel the mud into the form, overfilling each compartment slightly; spread and push the mixture into the corners with your hands *(left)*.
◆ Tamp the mud down hard with the end of a 2-by-4, then wet the 2-by-4 and pull it over the form edges to level the adobe.
◆ Lift off the form immediately, and clean it well with water and a stiff brush.
◆ Repeat the molding procedure until you have used up the prepared mud.
◆ Allow the bricks to dry flat until they are firm enough to hold their shape—one to four days, depending on air temperature and humidity.
◆ Stand the bricks on edge to dry both faces; in six weeks they will be ready for use or storage; store them in rows slightly tilted against a central pillar of flat bricks and cover them with plastic.

# Paving with Soil-Cement

Soil-cement, an alternative to concrete paving, combines Portland cement with soil to create a durable, weather-resistant surface. Although it is not as strong as concrete, it is far easier and cheaper to build with. Soil-cement looks more like dirt than concrete; it has a rough, often crackled finish and an earthen hue, and works best for informal walkways, as a subpavement for a brick or tile patio, or as a driveway. For foot traffic and subpavement, you will need 4 inches of soil-cement; for a heavily used driveway or any surface subject to freezing and thawing 5 to 6 inches is recommended.

**Calculating Proportions:** The durability of soil-cement depends on using the correct ratio of cement to soil, which in turn depends on the nature of the soil. The correct ratio ranges from about 10 percent cement for sandy, gravelly soils to 16 percent for soils with some organic matter. Sand will need to be added to soils that are rich in clay or organic matter. To find the best proportions of water, soil, and cement for your project, make test samples as you would for adobe bricks *(page 63)*.

To determine the total amount of cement you will need, multiply the percentage of cement in the best sample by the total volume of the job—if you have determined that your soil will need 10 percent cement and you plan to cover a pathway 20 feet long and 3 feet wide with 4 inches of soil-cement, the total soil volume is 20 cubic feet, and the cement needed will be 2 cubic feet, or two 94-pound bags.

**Work Strategies:** Cement begins to set up as soon as it is spread on the soil, so work with only as much cement as you can incorporate, moisten, and compact in one work session. Divide a large site into sections, each containing the correct amount of soil for one bag of cement; mark off the sections with chalk or string. Complete one section at a time, but do not wait any longer than 24 hours to do the next section.

### TOOLS

| | |
|---|---|
| Shovel | Hoe |
| Square shovel | Handsaw |
| Maul | 4 x 4 tamper |
| Rotary tiller | Screwdriver |
| Garden rake | Lawn roller |

### MATERIALS

| | |
|---|---|
| Lumber (2") | Door pulls |
| 2 x 2s, 2 x 4 | Polyethylene sheeting (6-mil) |
| Portland cement | |

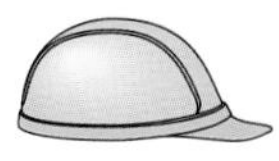

### SAFETY TIPS

*Put on steel-toed boots to operate a rotary tiller, and wear gloves when working with cement.*

### A ROTARY TILLER FOR CUTTING THROUGH SOIL

Breaking and turning over soil is hard work, but it can be made easier with a rotary tiller. The tiller can also be applied to the task of incorporating cement into the soil. It works much like a lawn mower, cutting as it moves forward—slowly through hard-packed earth, and more quickly through loose soil. A drag bar controls tilling depth as well as the forward movement of the machine, and can be adjusted to suit the type of soil being tilled. A rear guard, in addition to being a safety feature, helps break up the larger chunks of tilled soil and smooth the surface.

**1. Preparing the soil.**

◆ Remove any sod or topsoil covering the paving site.

◆ Brace the surrounding earth with 2-inch-thick boards held in place by 2-by-2 stakes, or create a permanent border of bricks.

◆ With a shovel or rotary tiller, turn over and pulverize the remaining soil to the planned depth of the paving, breaking up any large clods of dirt and removing any stones that are more than 1 inch in diameter *(right).*

◆ With a garden rake, redistribute the soil to form a smooth layer of uniform thickness.

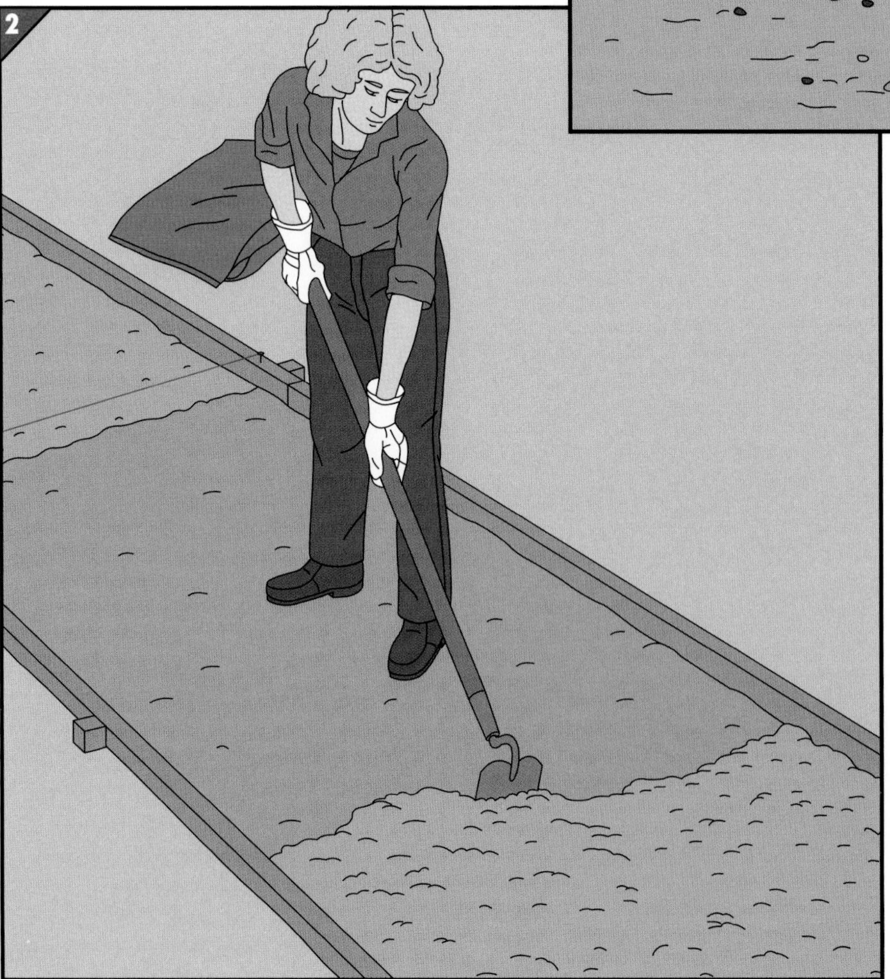


**2. Incorporating the cement.**

◆ Working in sections, spread the predetermined percentage of Portland cement in an even layer over the soil *(left).*

◆ Work in the cement with a hoe or rotary tiller. Periodically cut holes through the mixture with a shovel to see if the color changes, and continue to till the mixture of soil and cement until it is a uniform color throughout the depth planned for paving.

◆ When the mixture is uniform, rake the surface smooth.

### 3. Adding water.

◆ Lightly moisten the surface of the soil-and-cement mixture with a fine spray of water from a garden hose, dampening but not soaking it *(left)*.
◆ Work the water into the mixture with a hoe or rotary tiller to the full depth of the paving *(Step 2)*.
◆ Test the mixture for its moisture content by squeezing a handful; it is sufficiently moist when the handful forms a ball that can be broken apart cleanly, without crumbling. Add more water if necessary, work it into the mixture, and test again.
◆ When the correct moisture content is reached, rake the surface smooth.

### 4. Compacting the surface.

◆ Make a tamper by cutting a 4-by-4 waist high and then screwing a pair of large door pulls to the sides.
◆ Pound the soil-cement with the tamper, using enough force to compact it to its full depth *(right)*.
◆ When the paving is evenly tamped, score it lightly with the rake to remove irregularities.
◆ Cut a 2-by-4 to the width of the paving, then pull it over the surface to smooth it.
◆ Dampen the surface lightly with a garden hose, then tamp it again, so that the surface is as firm as hard-packed earth.
◆ For a very smooth finish, run a lawn roller or a tennis-court roller over the soil-cement.
◆ If you are working in sections, proceed to the next section, tilling 2 inches into the border of the finished area to link the sections.
◆ Cover the finished area with 6-mil polyethylene and let it cure for three days.

# 3 Walls of Brick, Block, and Stone

Whether you are building a concrete-block structure, raising a decorative brick wall, or putting up a masonry arch, planning and accuracy are essential to the success of the project. Even a stone wall, with its seemingly random pattern, requires careful selection of stones and precision in placing them. Beyond standard masonry skills, you will need to use a few special techniques to build these projects.



Anchoring brick veneer with a wall tie →

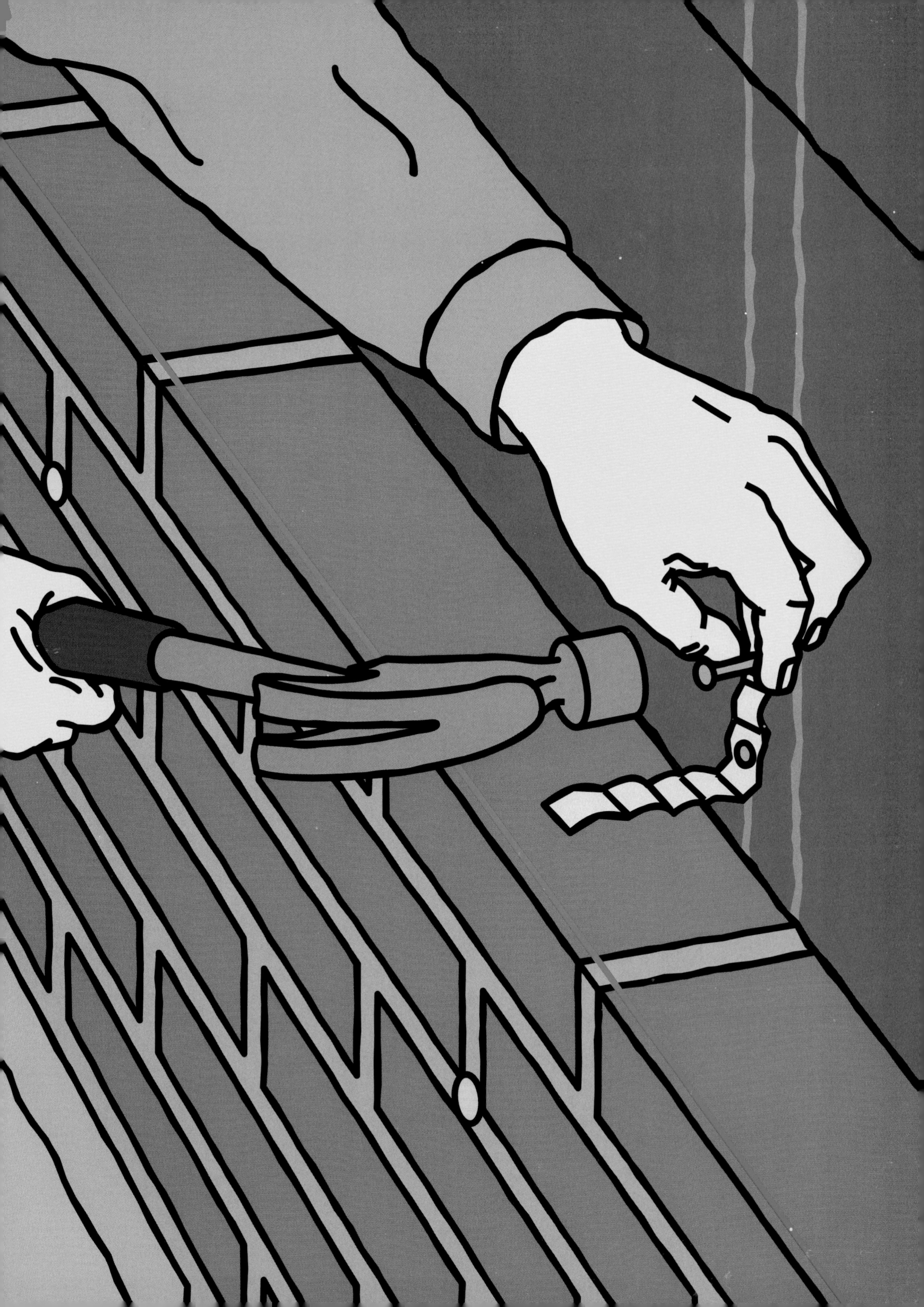

# Building with Concrete Block

Concrete blocks provide more strength and durability for the time and labor expended than virtually any other masonry material. They are ideal for structures such as basement walls, or even a small building such as a shed or garage.

To help make construction quicker, many structural materials are sized to match the dimensions of blocks. Precast lintels, needed to support the weight of the wall over door or window openings, are available in widths that conform to block widths.

**Planning a Building:** Putting up an entire building of concrete blocks requires careful planning. In a region with seismic activity or high winds, the walls are reinforced with rebar *(pages 74-75 and 80)* and require blocks without end flanges *(page 10)*. A long wall will need a control joint *(page 78)* to confine any cracking that does occur to this deliberately weakened point. If the structure will have interior walls, windows, doors, or floor joists *(pages 76-79)*, you will have to build them as you raise the outer wall.

**Storing Blocks:** Since individual blocks weigh between 25 and 35 pounds, it is important to have them delivered as close to the work site as possible. Plan a storage area easily accessible to the delivery truck. An ideal spot is on level ground, about 6 feet inside the footing. Cover the stored blocks with tarpaulins to keep them dry, since wet blocks expand slightly and may cause cracks in mortar joints when they dry and shrink.

**TOOLS**

- Mason's trowel
- Level (4')
- Mason's line
- Bolt cutters
- Rebar cutter
- Rebar bender
- Brickset
- Ball-peen hammer
- S-shaped jointing tool (6")
- Jointer (22")
- Stiff-bristle brush
- Raking tool
- Tin snips
- Circular saw
- Electric drill
- Crowbar
- Caulking gun

**MATERIALS**

- 2 x 4s
- Joist lumber
- Pressure-treated 2 x 8s
- Lumber for story pole
- Common nails ($2\frac{1}{2}$", $3\frac{1}{2}$")
- Lag screws ($\frac{1}{2}$" x 4") and shields
- Screws ($\frac{1}{2}$" x 4") and toggles
- Anchor bolts
- Mortar and grout
- Concrete blocks and lintels
- Rebar, supports, and tie wire
- Steel joint-reinforcement wire
- Bricks
- Wire mesh
- Perforated metal strapping
- Building paper
- Flexible caulk
- Neoprene flange

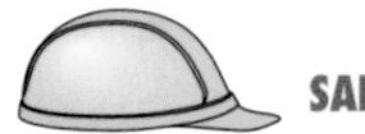

**SAFETY TIPS**

*Protect your hands with gloves when working with mortar or handling concrete block, and put on goggles when hammering or operating power tools. Wear steel-toed boots to prevent injury from dropped blocks.*

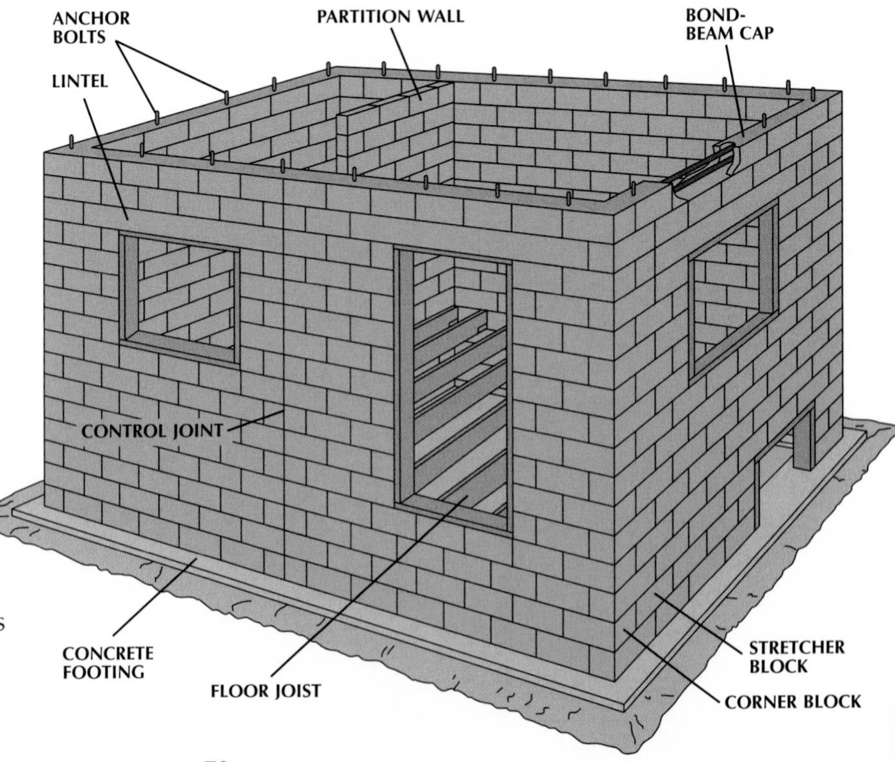


**Anatomy of a block building.**
This building's walls rest on a concrete footing, and every second course has steel joint-reinforcement wire laid horizontally to help prevent cracking *(page 74)*. On walls longer than 30 feet, a vertical control joint is placed at the end of a lintel above a window or door openings as close to the middle of the wall as possible to absorb expansion and contraction during temperature changes *(page 78)*. Precast concrete lintels lend support to courses above door and window openings *(page 79)*. Wood floor joists are set into pockets *(page 77)*, and an interior partition wall is locked into the exterior wall in a toothed pattern *(pages 76-77)*. The roofline is strengthened laterally with a cap of bond-beam blocks in which horizontal rebars have been laid and covered with grout *(pages 79-80)*. This cap is also the base for the L-shaped anchor bolts that tie the roof to the wall.

# LAYING THE FIRST COURSE

**1. Mortaring the footing.**

◆ After building a footing suitable for the planned structure *(pages 44-50)*, stretch strings between the batter boards at the outer building lines.

◆ Lay out a dry run of blocks to determine their correct positions.

◆ At one corner of the building lines, trowel a level bed of mortar about $1\frac{1}{2}$ inches thick and 2 inches wider than the width of the block; spread it only far enough to reach 1 inch beyond each end of the block *(above)*.

◆ Set a corner block in the mortar, with the thicker webs facing upward; allow the weight of the block to compress the joint to a thickness of $\frac{3}{8}$ inch.

◆ Spread mortar and set another corner block at the opposite corner.

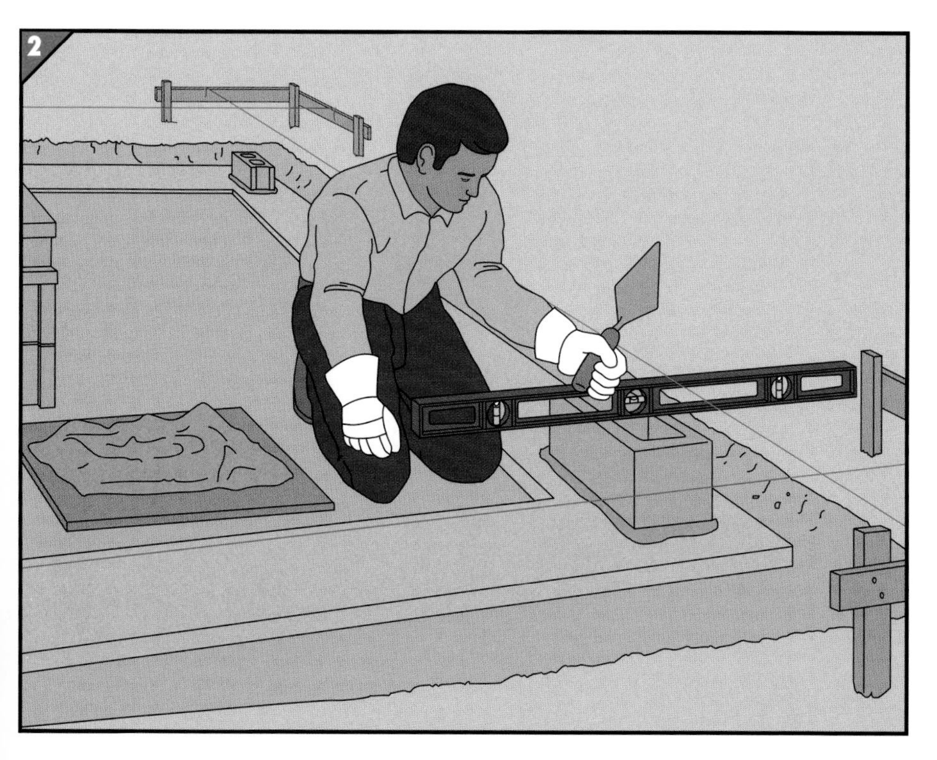


**2. Leveling the corner blocks.**

◆ Rest a level on a corner block, and tap the block with the handle of the trowel until the block rests level on its mortar bed *(left)*.

◆ Level the other corner block in the same way.

◆ Once the two corner blocks are mortared and leveled, string a mason's line along their outside faces, to serve as a guide in aligning the first course of the wall.

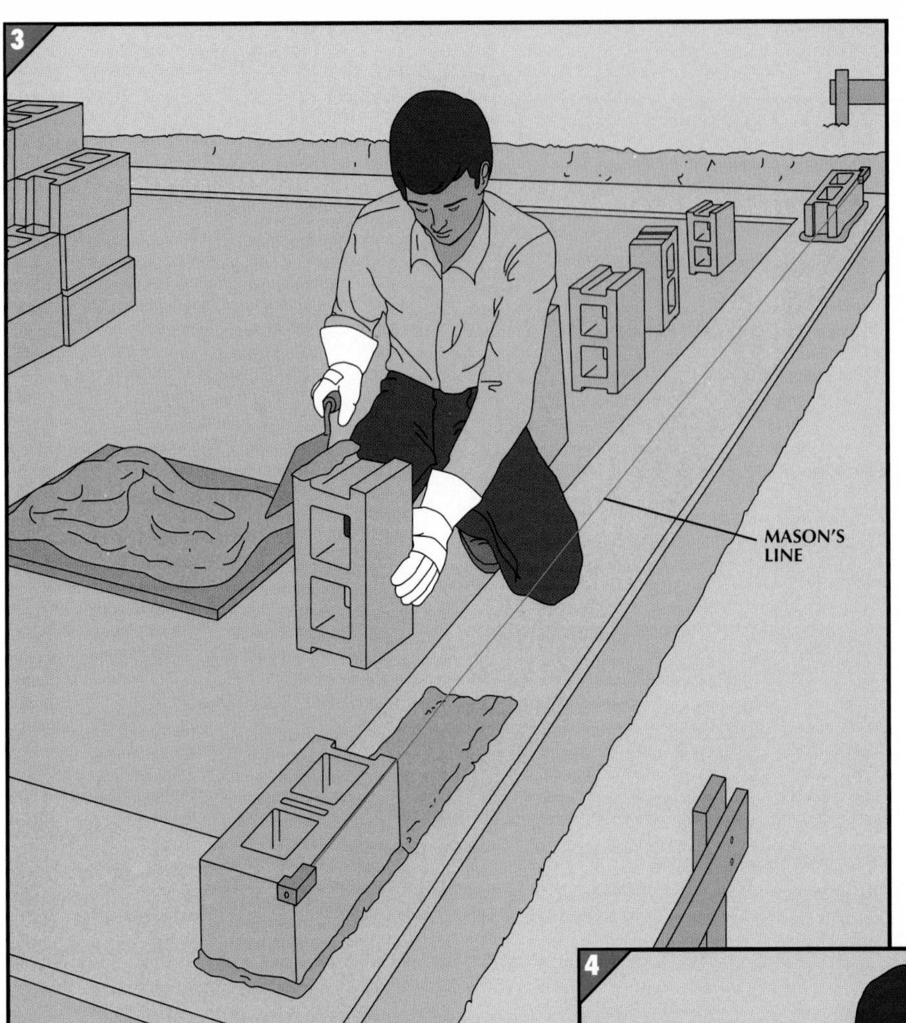


### 3. Buttering blocks.

◆ Turn a stretcher block on end, then with the bottom of the trowel press a thick layer of mortar onto each flange; work the mortar to form a peak about 1 inch high *(left).*
◆ With the thicker webs of the block facing upward, press the mortared flanges of the stretcher block against the flanged end of the corner block to form a masonry joint $\frac{3}{4}$ inch wide; ensure that the block lines up with the mason's line.
◆ Continue to lay succeeding stretcher blocks, substituting a 4-inch-wide partition block wherever a planned interior wall will intersect with the outer wall *(pages 76-77),* until you reach the center. Lay blocks from the opposite end of the wall until there is room left for only one block. If the wall requires a control joint, begin it in this course *(page 78).*
◆ Butter the flanges at the ends of the two blocks forming the opening, as well as the flanges of this last, closure piece.
◆ Lower the closure piece into the wall.
◆ Complete the first course of the remaining walls in the same way.

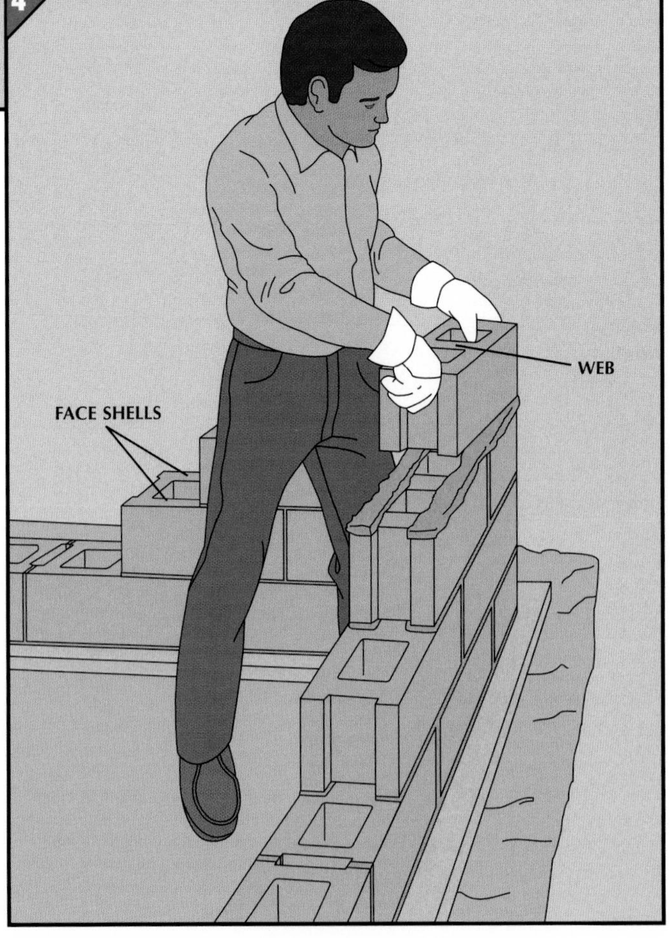


### 4. Building up a lead.

◆ At each corner, build a stepped second, third, and fourth course, called a lead *(right),* mortaring only the face shells—and not the webs—of the blocks in the underlying course, and staggering the position of the vertical joints. Set joint-reinforcement wire in place at every other course *(page 74).*
◆ Complete the second course of all the walls, using the mason's line for alignment; work from the ends to the center and locate the closure piece in a different spot than that of the first course.
◆ Complete the third and fourth courses all the way around in the same way.
◆ Continue the wall upward in the same way, building corner leads for every three courses, and periodically measuring the courses with a story pole, and checking for plumb and level *(pages 73-74).*

# CHECKING FOR PLUMB AND LEVEL

**Measuring with a story pole.**

◆ Make a story pole by drawing course lines with a felt-tipped pen along a straight piece of lumber, marking off the actual height of a block, then the thickness of the mortar joint, then another block height and mortar-joint thickness, and so on along the length of the wood. A mason's ruler can be used instead *(photograph)*.

◆ Beginning at the leads, check with the story pole that the top of each block course corresponds to the course lines, and that the mortar joints are the correct thickness *(right)*.

◆ If the mortar joints are too thick, tap down on the blocks gently with the handle of the mason's trowel; if they are too thin, lift the block, add mortar, and lay the block again.

◆ Continue checking at about 4-foot intervals to make sure each course is rising evenly.

**Plumbing and leveling.**

◆ Before the mortar sets, hold a 4-foot level vertically against three or more courses *(above, left)*.

◆ Nudge misaligned blocks into place with the trowel handle, tapping gently to avoid breaking the masonry bonds.

◆ Lay the level on top of a single course *(above, right)*, and make any necessary adjustments with the trowel handle.

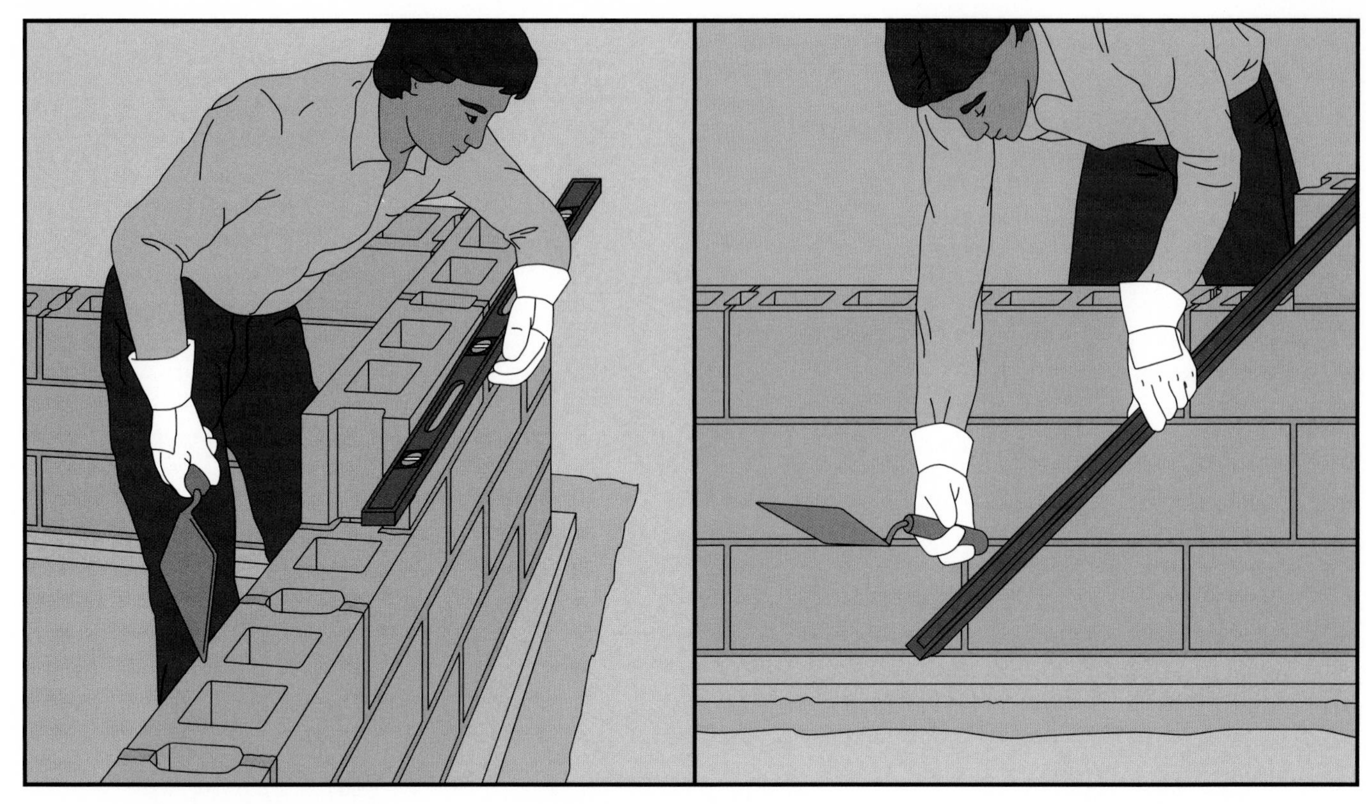

**Checking for bulges.**
◆ With a level serving as a straightedge, hold the level horizontally along the outside face of each course to check the wall for alignment *(above, left)*.
◆ Tap gently with your trowel handle to realign any mislaid blocks.
◆ Next, hold the level diagonally along the descending corners of three or more courses *(above, right)*, and realign as necessary with the trowel handle.
◆ Check once again for plumb and level *(page 73, bottom)*.

## REINFORCING WALLS

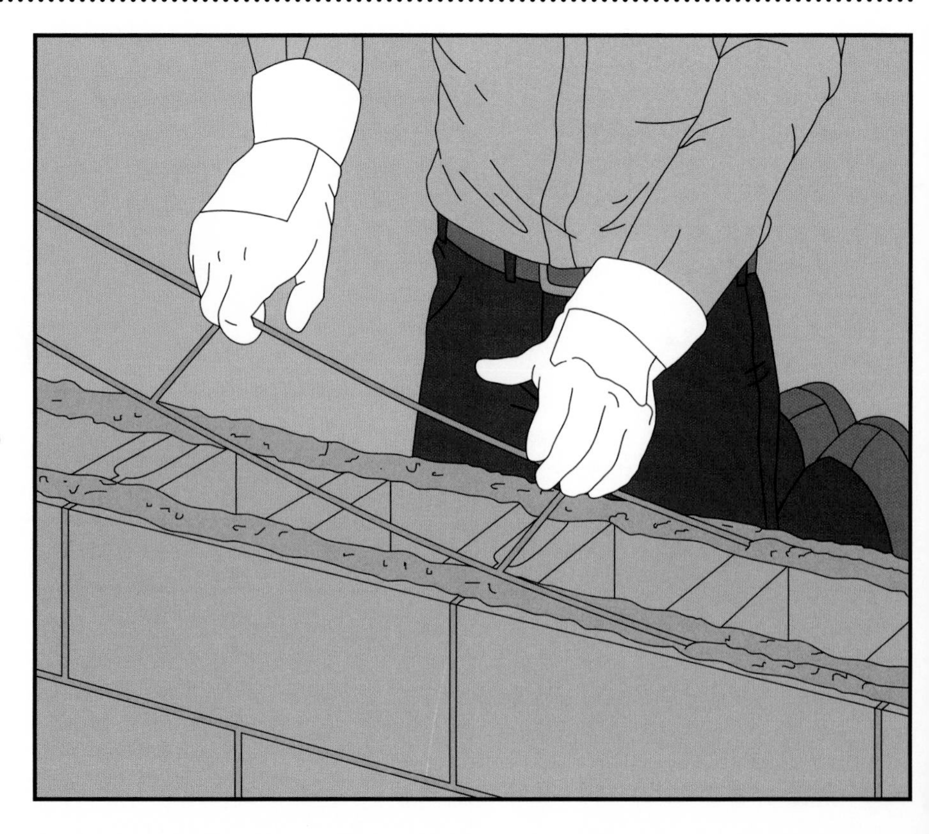

**Horizontal wire grid.**
◆ Lay sections of steel joint-reinforcement wire along the mortared tops of every other course of blocks *(right)*. Where necessary, cut the wire grid to the correct length with bolt cutters.
◆ Where two lengths of reinforcement join, at corners or along a wall, overlap the two lengths of wire by the width of a block.

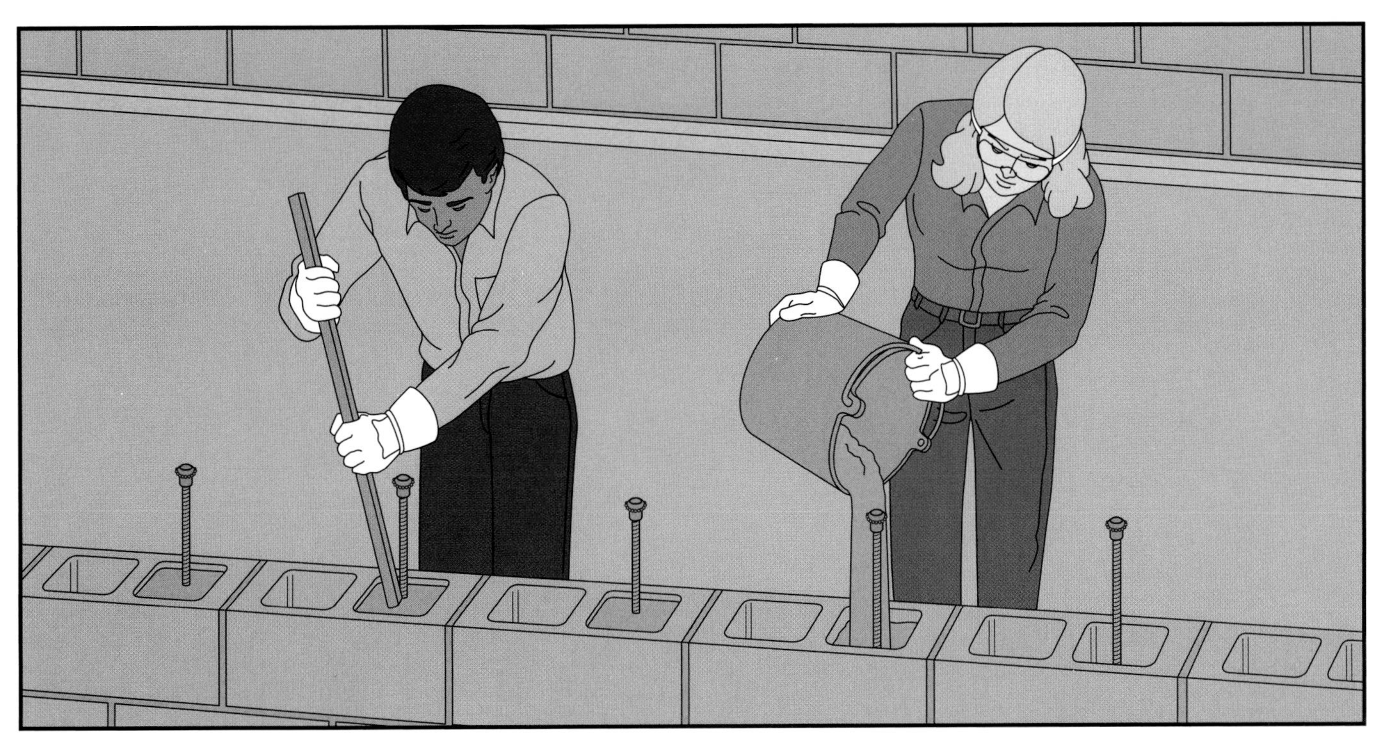

**Vertical reinforcement.**
◆ At each point where vertical rebar rises from the footing *(page 48)*, lower buttered blocks over the rebar and set them in place *(page 72, Step 3)*. Mortar the webs as well as the shell-faces of the cores containing vertical reinforcement.
◆ After 3 courses, fasten additional 4-foot lengths of rebar to the footing rebar with tie wire, overlapping the bars by 24 inches. Continue to build up the wall, lifting the blocks over the rebar.
◆ When you have completed the sixth course, add 4-foot lengths of rebar, tying them to the rebar already in place.
◆ At the sixth course, pour grout into each cavity with vertical rebar, stopping about 1 inch from the top. Have a helper push down the grout with a stick to ensure that all the cores are filled *(above)*. After 10 minutes, have your helper push down the grout again.
◆ Continue to build the wall upward, adding rebar every three courses and grouting every six courses.
◆ Top the wall with a course of special bond-beam blocks with cores *(page 80)*.

## CUTTING BLOCK

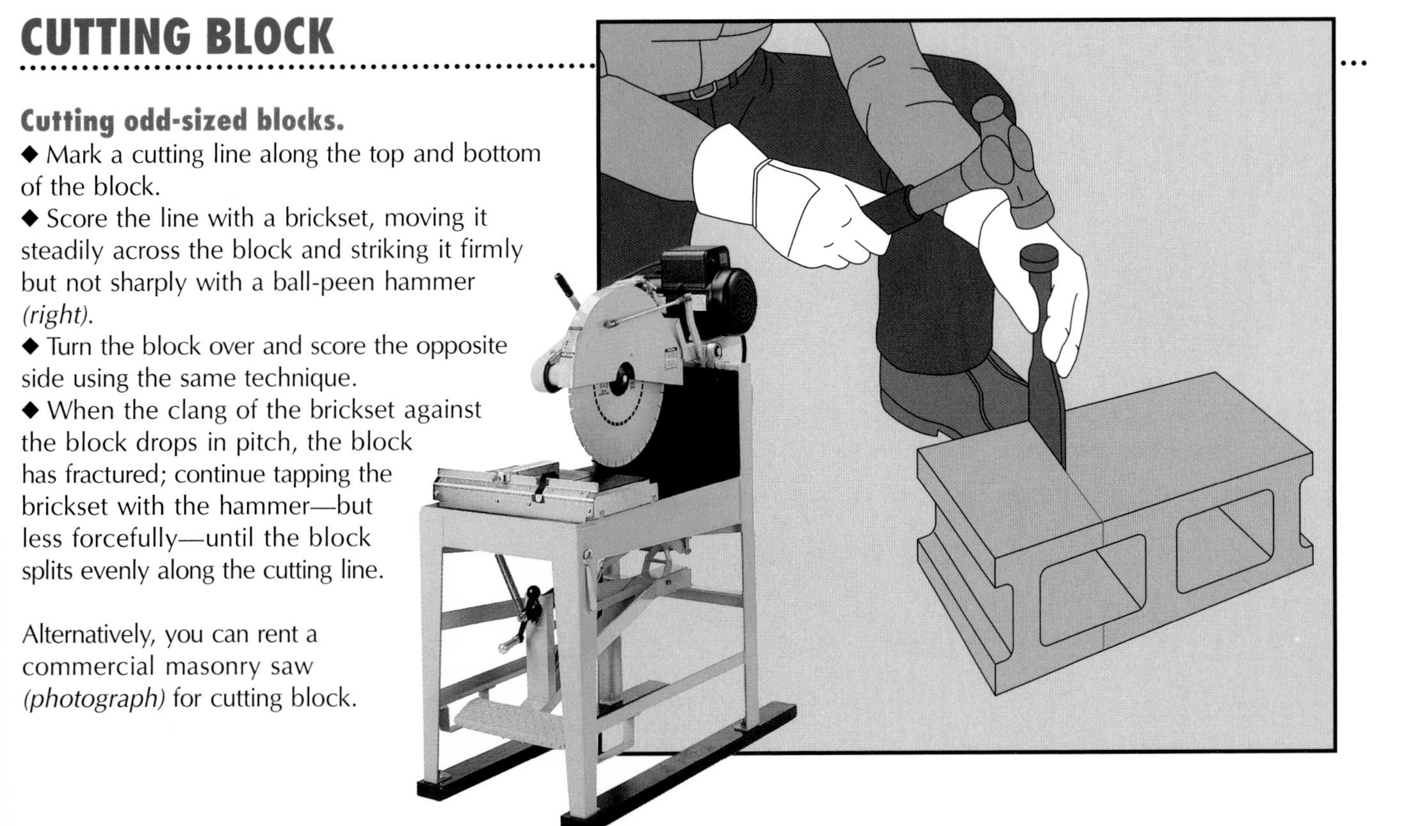

**Cutting odd-sized blocks.**
◆ Mark a cutting line along the top and bottom of the block.
◆ Score the line with a brickset, moving it steadily across the block and striking it firmly but not sharply with a ball-peen hammer *(right)*.
◆ Turn the block over and score the opposite side using the same technique.
◆ When the clang of the brickset against the block drops in pitch, the block has fractured; continue tapping the brickset with the hammer—but less forcefully—until the block splits evenly along the cutting line.

Alternatively, you can rent a commercial masonry saw *(photograph)* for cutting block.

## TOOLING JOINTS

**Tooling the mortar.**

◆ When the mortar is dry enough that it will not adhere to your thumb when you press it against a joint, compact the vertical joints with the convex face of a 6-inch S-shaped jointing tool *(above, left)*, applying enough pressure to seal hairline cracks caused by the shrinkage of drying mortar.

◆ Finish the horizontal joints with a 22-inch jointer *(above, right)* in the same way.

◆ Trim off any remaining mortar burrs with the point of a trowel.

◆ When the joints are completely dry—after about 24 hours—groom the wall with a stiff-bristle brush to remove loose particles of mortar.

## INTERSECTING WALLS

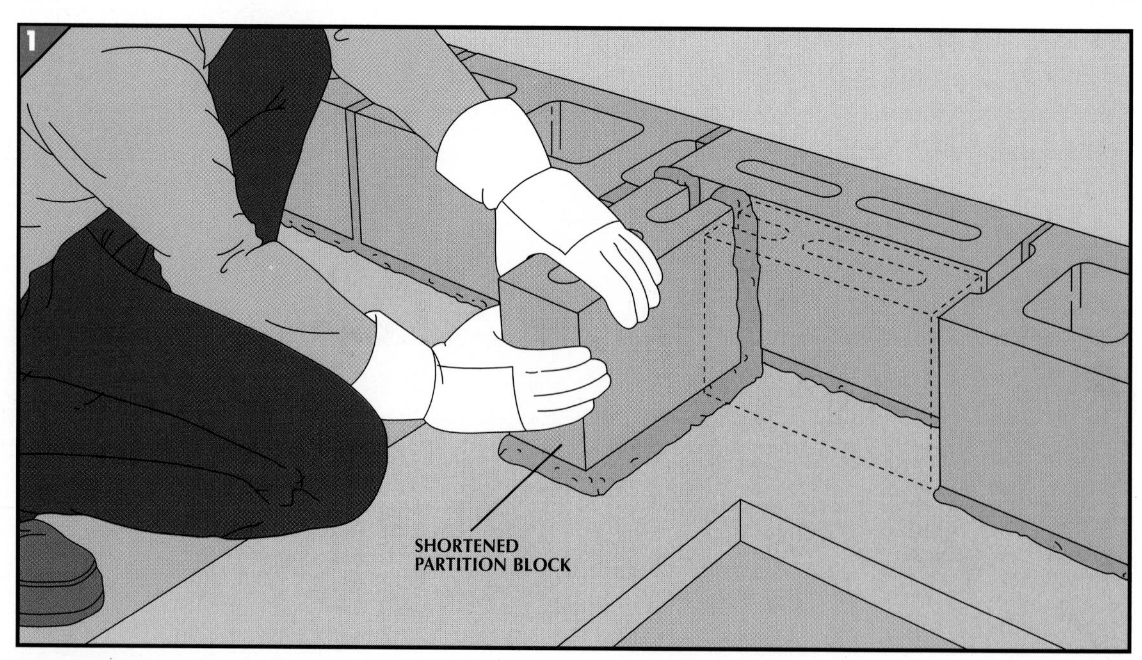


**1. Beginning the pattern.**

◆ In the pocket created by the initial partition block set in the outer wall *(page 72, Step 3)*, set another partition block, shortened by 4 inches, perpendicular to the outer wall *(above)*.

◆ Fill in the remaining space along the outer wall with a partition block that has been cut 12 inches long, restoring the wall to its full depth *(dashed lines)*.

**2. Toothing the wall.**

◆ In the second course, lay standard stretcher blocks along the outer wall.
◆ For the second course of the interior wall, lay a full-sized partition block over the shortened partition block on the first course.
◆ Prop up the overhanging block with an upright 8-inch brick, mortared at both ends, or an 8-inch scrap of concrete.
◆ Alternate these two courses as you raise the outer wall, creating a toothed joint for the interior wall.
◆ After the mortar joints of the toothed blocks have set, remove the props by twisting them out—tap them lightly if necessary—and complete the intersecting wall.

## FLOORS, DOORS, AND WINDOWS

**Securing floor joists.**

◆ Determine the positions of the joists and, as you are erecting the side walls, lay wire mesh beneath the course of standard blocks that the joists will be resting on.
◆ On top of this course lay partition blocks, creating a 4-inch-deep ledge.
◆ Cut 12-inch lengths of perforated metal strapping to make joist anchors, curving each to 90 degrees and twisting one end so that it can fasten to a joist.
◆ At every joist position, fill the core of the underlying block with grout, then install a joist anchor, setting the end of the anchor into the grout.
◆ Cut joists to length, then cut both ends of each joist at an angle, so that the top is 2 inches shorter at each end than the bottom.
◆ Position the joist on the ledge, and fasten the anchors to the sides of the joist with $2\frac{1}{2}$-inch common nails *(right)*, clinching the nails on the other side by hammering the ends flat against the joist.
◆ Fill the spaces between joists with partition blocks cut $13\frac{1}{2}$ inches long, leaving a $\frac{1}{2}$-inch space on both sides of each joist to keep moisture from collecting.

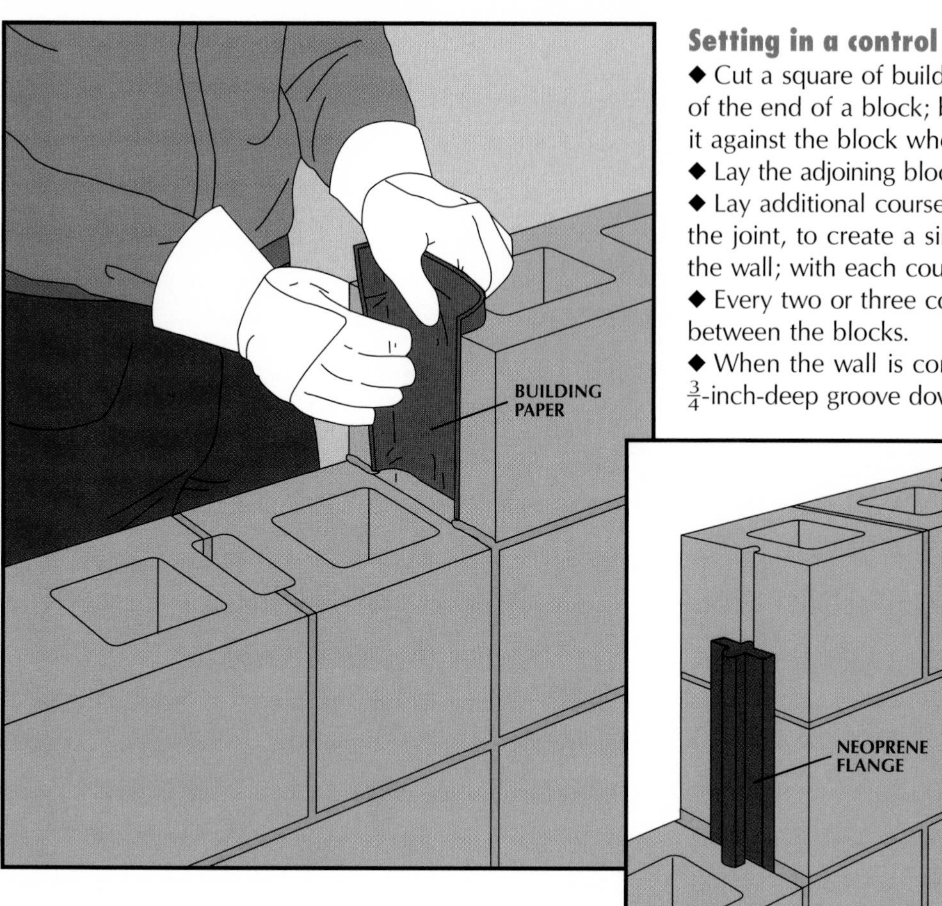


**Setting in a control joint.**

◆ Cut a square of building paper to fit within the contours of the end of a block; beginning at the first course, press it against the block where the control joint will be.

◆ Lay the adjoining block in the usual way *(page 72, Step 3)*.

◆ Lay additional courses, alternating full and half blocks at the joint, to create a single continuous joint the height of the wall; with each course, press in building paper *(left)*.

◆ Every two or three courses, pour grout into the cavity between the blocks.

◆ When the wall is complete, use a raking tool to form a $\frac{3}{4}$-inch-deep groove down the full length of the control joint.

◆ Pack the groove with flexible caulk to weatherproof the joint.

For blocks without flanges, an alternate seal called a neoprene flange, available from block manufacturers, slips into the sash groove in a block specially made with this slot *(inset)*; the flange serves the same purpose as building paper, and the joint is mortared and caulked in the same way.

**Door and window frames.**

◆ Lay wire mesh beneath the course directly below the planned door or window. When the course is in place, fill the cores of the blocks with grout.

◆ When the block wall reaches the height of the top of the planned opening, cut two pressure-treated 2-by-8s as long as the opening is wide. Position one as a bottom frame, drilling through the board and into the grout near each side of the opening and at 2-foot intervals in between for lag shields and $\frac{1}{2}$- by 4-inch lag screws, then fasten it in place.

◆ Cut two more 2-by-8s to fit on the bottom frame that end $1\frac{1}{2}$ inches below the top of the opening. Set them in place and brace them with scrap lumber, then drill holes at 2-foot intervals for $\frac{1}{2}$- by 4-inch screws and metal toggles, and fasten the boards to the blocks *(right)*.

◆ Secure the remaining 2-by-8 to the top of the side pieces with $3\frac{1}{2}$-inch common nails.

◆ Install a lintel over the opening *(opposite)*.

◆ After the wall is complete, caulk the joints between the frame and the wall.

Metal door and window frames come with projecting anchors, which are mortared into place to secure the frame as the wall is built.

**Raising a lintel.**

◆ Lay the course above the door or window, leaving an opening for the lintel.
◆ To build a sling for the lintel, drill a hole at each end of a 2-by-4 that is 20 inches shorter than the length of the lintel, then run a length of rope through each hole.
◆ Set the lintel on the sling and make sure the rope loops are equal in size so the lintel will be balanced.
◆ Remove the guard rails from the scaffold temporarily, then with two people on the scaffold, have a third person hoist the lintel *(left)*. Replace the guard rails.
◆ Mortar the ends of the lintel, then maneuver it into place on the wall by levering it off the sling with a crowbar.

## FINISHING THE TOP OF THE WALL

**1. Laying the top course.**

◆ Place wire mesh across the cores of the corner blocks in the second-to-last course. If the bond-beam blocks have cores *(page 80, box)*, lay mesh all along the second-to-last course.
◆ Notch the partition walls of the corner blocks with a brickset to make the channel continuous *(inset)*, then set them in place over the wire mesh.
◆ Lay bond-beam blocks for the rest of the top course *(right)*.

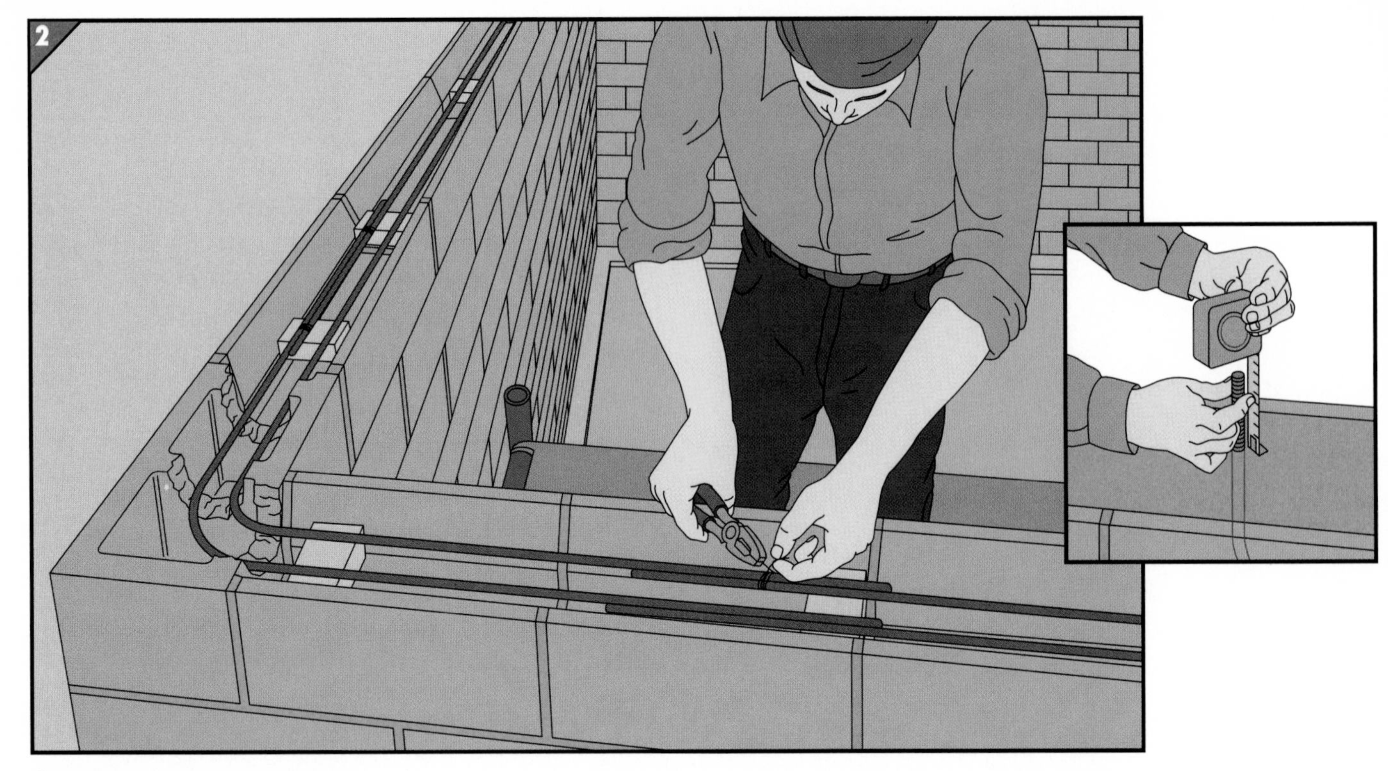

**2. Adding reinforcement.**

◆ Set $\frac{1}{2}$-inch rebar—or the size specified by the local building code—into the channel in the bond-beam blocks in parallel rows 2 inches apart and resting on concrete supports. Overlap lengths of rebar 12 inches and tie them together with wire *(above)*; at corners, bend the rebar into a 90-degree angle *(page 48)*.

◆ Pour grout into the channel in the blocks, filling to the top of the blocks. Use a stick to make sure that grout fills the area around and beneath the rebar.

◆ With a 2-by-4, strike off the grout level with the tops of the blocks.

◆ Set anchor bolts, of a size prescribed by code, in the grout, spacing them according to code and centering them in the block so they protrude $2\frac{1}{2}$ to 3 inches *(inset)*.

## CAPPING A REINFORCED BLOCK STRUCTURE

In areas with seismic activity or high winds, the vertical rebar that was cast into the footing *(page 48)* and extended upward *(page 75)*, is tied to the horizontal rebar in the bond-beam cap. To allow the vertical rebar to reach to the top of the wall, bond-beam blocks with cores are used in place of channel-shaped bond-beam blocks. Mesh is laid all along the top of the final course of standard blocks—except for over the cores containing vertical rebar. The vertical rebar is then cut 1 inch below the tops of the bond-beam blocks and tied to the horizontal rebar. The bond-beam blocks are filled with grout, which is leveled off to the height of the blocks.

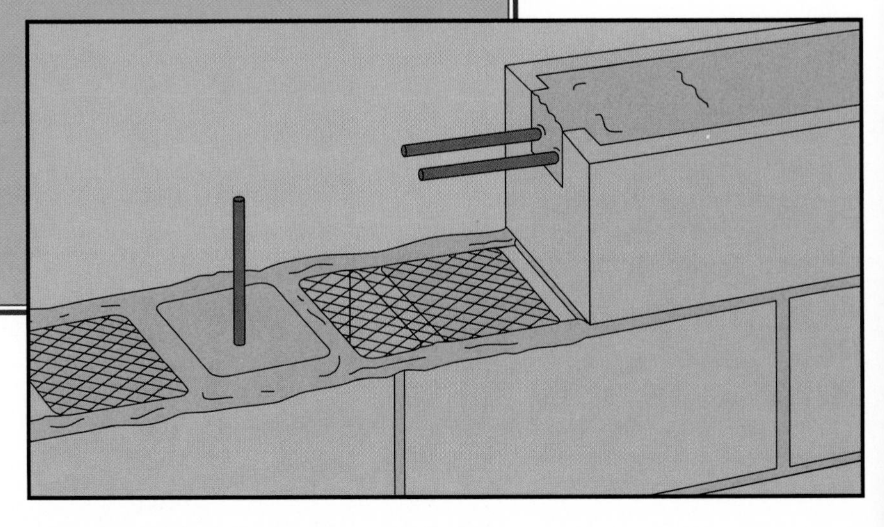

# Veneering a House with Brick

Building an outer wall of brick over an inner wood frame is far less costly than constructing a solid brick wall, but yields many of its advantages: Brick is more resistant to weather than wood siding, and it requires very little maintenance.

**Footing, Foundation, and Eaves:** You can also veneer a house that is already covered with siding—if the building meets certain conditions. The footing must be wide enough to accommodate the extra layer of brickwork. And if the house has a basement, you will need to build a foundation wall of 4-inch concrete blocks to support the bricks. If the footing extends 5 inches beyond the foundation, it is ample for the new wall; otherwise consult a specialist about having the footing extended. You may need to extend the overhang of the eaves as well to accommodate the thicker wall. Submit a plan for the project to the local building department for approval.

**Preparatory Work:** You will need to cover the existing siding with building paper, extend the window and door casings to be deep enough to meet the new brickwork, and locate wall studs for attaching the wall ties that will anchor the brickwork to the existing walls. One wall tie is set for every 12 bricks, although on gables you will need one for every six bricks.

The final step in planning is to plot the placement of courses, which may have to be adjusted vertically in order to avoid splitting bricks horizontally as you build around windows and doors and under the soffit at the eaves *(page 82)*. If you plan to enliven the brickwork with quoins, decorative patterns, or arches *(pages 99-102 and 112-120)*, draw a section of the wall on graph paper.

Rent scaffolding, and plan to do your bricklaying in good weather, when neither rain nor freezing temperatures will adversely affect the mortar.

**TOOLS**

- Level
- Hammer
- Mason's line
- Mason's trowel
- Jointing tool
- Jointer

**MATERIALS**

- 2 x 4s
- Common nails ($3\frac{1}{2}$")
- Galvanized common nails ($2\frac{1}{2}$")
- Roofing nails (1")
- Bricks
- Mortar
- Polyethylene sheeting (6-mil)
- Steel wall ties
- Rope or $\frac{3}{8}$-inch tubing
- Steel lintels ($\frac{1}{4}$" x $3\frac{1}{2}$" x $3\frac{1}{2}$")
- Molding

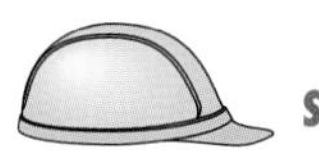

**SAFETY TIPS**

*Wear gloves and steel-toed shoes when handling bricks. Add goggles when working at eye level with mortar, or when you are driving nails.*

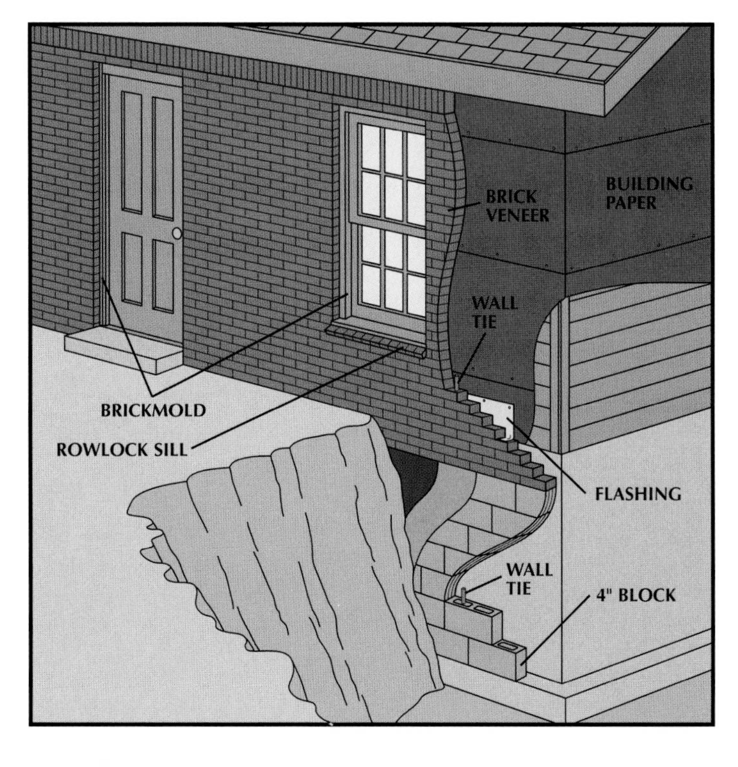


**Anatomy of a veneered house.**
The brick facing added to a wood-frame house rests on a 4-inch-wide concrete-block foundation built on the outer ledge of the house footing, but separated from the foundation by a 1-inch air space. At every two courses of block and at 6-foot intervals, corrugated-steel wall ties are anchored in the mortar joints and nailed to the existing foundation with $2\frac{1}{2}$-inch masonry nails. The block foundation rises to a point that will allow the first three courses of brick to be laid below grade, and is damp-proofed *(page 94)*.

Overlapping sheets of building paper protect the siding from moisture. The veneer bricks stand 1 inch away from the siding; wall ties anchor them every six courses to the wall studs. Steel lintels provide support over windows and doors. Moisture is directed out of the air space with flashing and rope wicks placed at the bottom of the veneer wall, above windows and doors, and below sills. Brickmold—wood molding 1 inch thick and $2\frac{1}{2}$ inches wide—is added to window and door frames. Below each existing window sill, rowlocks—bricks set on edge—create a second sill, angled to shed water.

# ALIGNING COURSES

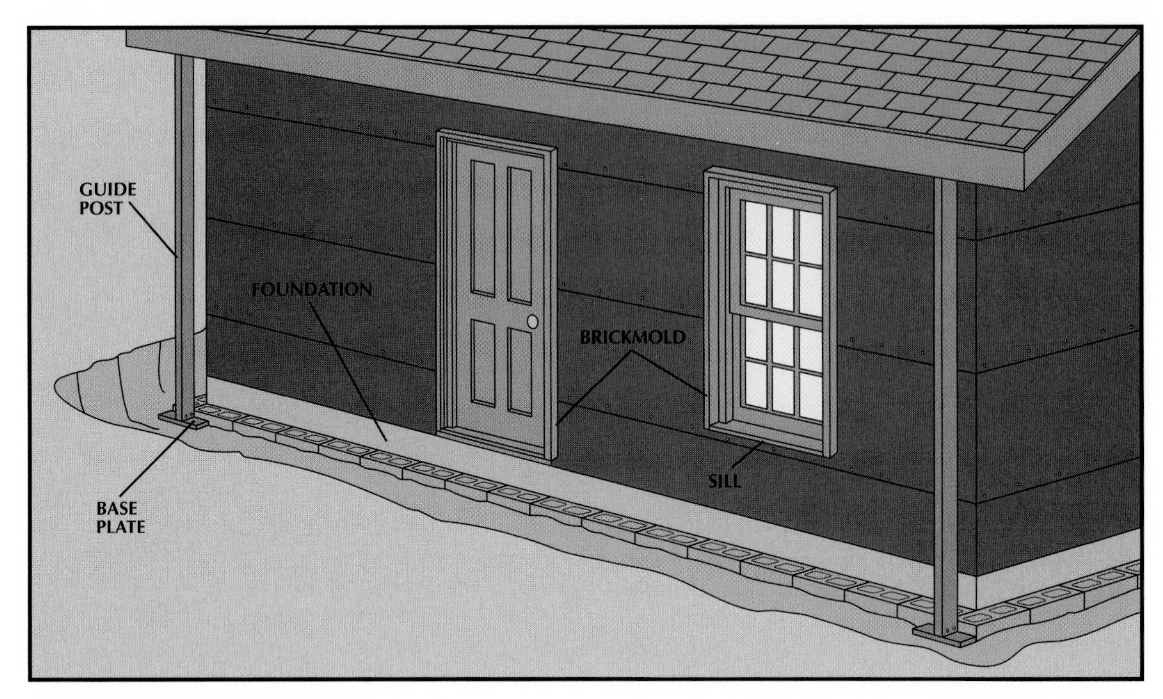


**A system for marking course levels.**
To keep from having to split bricks horizontally to fit at doors, windows, or the soffit, mark course heights in advance.

◆ Set two straight 2-by-4s, one at each end of the wall, on wood base plates so the distance between their inner faces and the wall is equal to the thickness of a brick, plus 1 inch for the air space between the wall and veneer.

◆ Plumb the posts and toenail them to the plates and the soffit, and mark on them a whole number of courses between the foundation and the bottom of the rowlock sills, which will extend $4\frac{1}{2}$ inches below the existing window sills. If sills are at different heights, choose one as a reference—the one at which split bricks would be most conspicuous. To arrive at whole-brick courses, you can vary the width of mortar joints from $\frac{1}{4}$ inch to $\frac{1}{2}$ inch. In the same way, measure and mark the courses between the bottom of the rowlock sills and the top of the wood brickmold on the windows and doors.

◆ Measure and mark the courses between the brickmold and the soffit.

# RAISING THE VENEER

**1. Laying the first course.**

◆ As shown in the inset, start at the right and lay a dry row of bricks along the top of the block foundation: Place the first brick 5 inches beyond the corner of the house, and adjust the width of the mortar joints so that the last brick—including the measurement for the mortar joint on the outside end—extends 1 inch beyond the far corner.

◆ Stretch a mason's line at the first-course markings on the guide posts.

◆ Set aside three or four bricks at a time, spread a bed of mortar on the foundation, then butter the ends of the bricks and position them in the mortar *(right).*

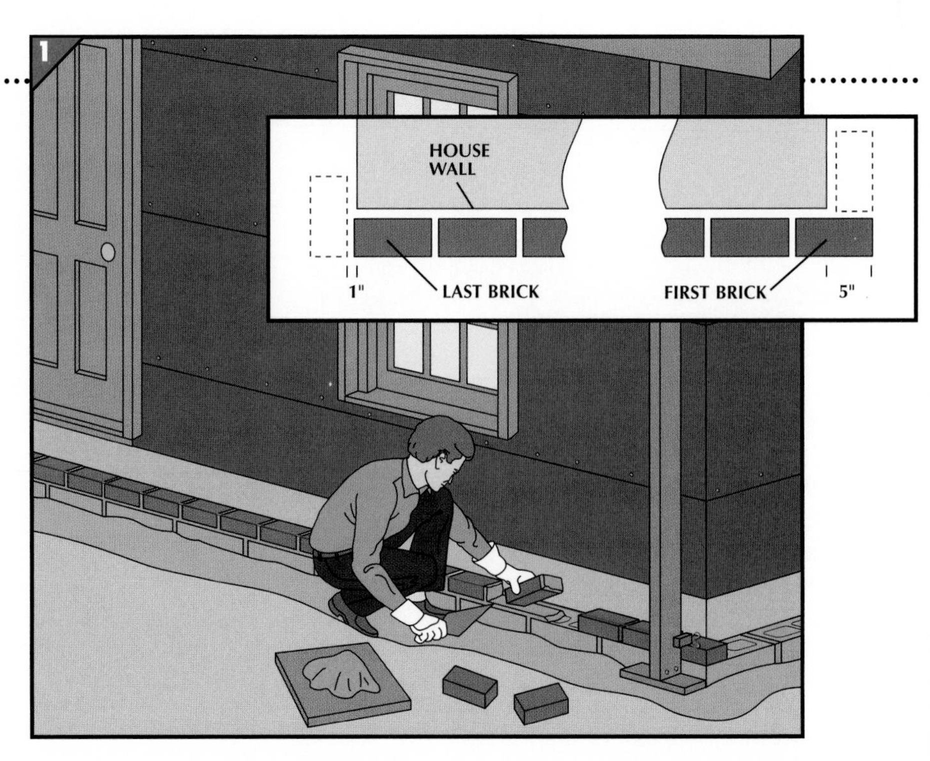

## A CONVENIENT BRICK TOTE

The task of shifting bricks from storage to work site can be made substantially easier with steel brick tongs *(photograph).* Place the tongs around the row of bricks you want to lift and raise the handle—the ends clamp tight. When you set the bricks down, the tongs release their hold. Depending on the model, the tool can be adjusted to accommodate different numbers of bricks—from as few as six to as many as eleven.

**2. Flashing the brickwork.**

◆ Add three courses of bricks in order to bring the top of the brick veneer just above ground level. For each course, align the bricks with the mason's line at the corresponding marks on the guide posts, and to allow for a doorstep do not place any bricks beneath the door opening; cut the bricks that will abut the step. As you work, use the trowel to remove any excess mortar that may have squeezed into the air space behind the bricks.

◆ Spread a 12-inch strip of 6-mil polyethylene sheeting along the top course of bricks and up the house wall, adjusting it so that the lower edge lies 1 inch from the outer face of the bricks; fasten the other edge to the wall with closely spaced 1-inch roofing nails *(above).*

◆ Spread a bed of mortar over the plastic and lay the next course of bricks as you did the last, but insert a piece of rope or $\frac{3}{8}$-inch tubing as long as a brick is wide at the base of every second mortar joint in this course *(inset)* to allow moisture to escape from the wall.

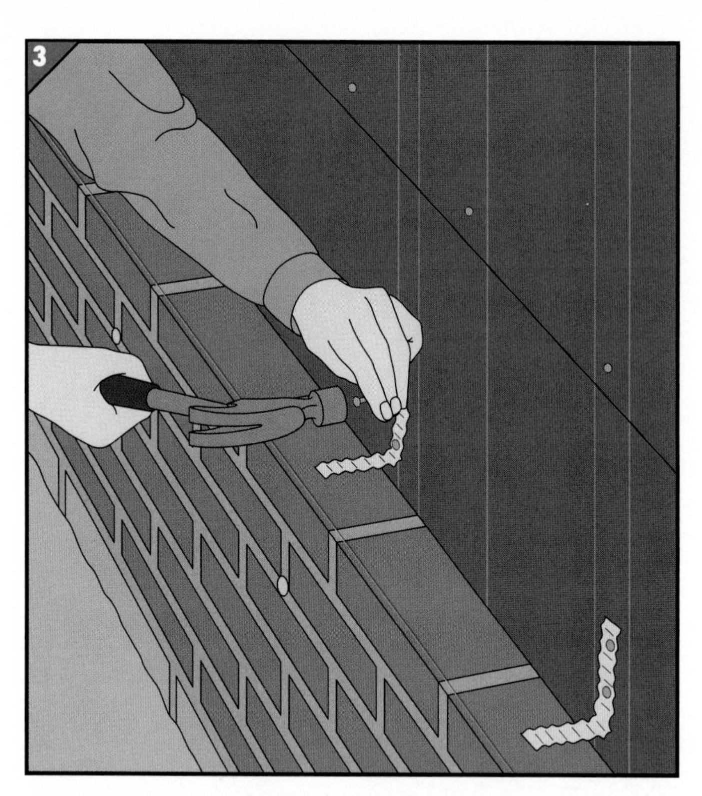

### 3. Anchoring the veneer.

◆ Moving up the mason's line on the guide posts as you go, lay subsequent courses of bricks with the vertical joints staggered and the ends of the courses forming a stepped diagonal away from the corners of the house. Cut any bricks abutting doors or windows to fit flush with the brickmold.
◆ When the veneer is six courses high, bend corrugated wall ties at their midpoint to form a right angle. Rest one leg of the tie on the brickwork and nail the other leg of the angle to a wall stud with $2\frac{1}{2}$-inch galvanized common nails *(left)*.
◆ Spread mortar and lay the next course of bricks, embedding the ties firmly in the masonry.
◆ For the adjacent wall, set up guide posts, marked as for the front wall, then lay the bricks and anchor them with wall ties in the same way as you did for the front wall.
◆ When you have covered about a third of the height of the wall, leaving a $4\frac{1}{2}$-inch gap beneath any windows *(page 85, Step 1)*, lay the corner sections of courses, wrapping each course around onto the adjacent wall as you work.
◆ Lay the adjacent corners of the other walls in this way, completing one-third of the height of each wall at a time and finishing the mortar joints *(page 76)* before moving to the next section.

## DEALING WITH WINDOWS AND DOORS

### 1. Positioning the lintel.

◆ Across the top of each door or window opening, lay a $\frac{1}{4}$- by $3\frac{1}{2}$- by $3\frac{1}{2}$-inch steel lintel that extends at least 8 inches beyond the opening on either side.
◆ Align the vertical flange of the lintel so that its front face is flush with the back face of the bricks on which it rests.
◆ Fill in the $\frac{1}{2}$-inch space between the front face of the bricks and the front edge of the lintel with a band of mortar, scraped from the back of the trowel *(above)*.

### 2. Setting bricks over the lintel.

◆ Cut a strip of 12-inch-wide 6-mil polyethylene sheeting the length of the lintel, then position and nail it *(page 83, Step 2)*.
◆ Adjust the mason's line and lay the next course of bricks; over the lintel, join the bricks end to end with mortar, omitting the bed of mortar beneath them *(above)*, and inserting lengths of rope or $\frac{3}{8}$-inch tubing at every second joint.

# CAPPING THE WALL

**Laying the final bricks.**

◆ When the veneer wall is three courses below the eave, set up the mason's line $\frac{1}{4}$ inch below the eave, and spread a mortar bed over a section of the last course of horizontal bricks.

◆ Butter one bed side of each brick for the final course and stand it on the mortar bed, lining up the top of the brick with the mason's line *(above).*

◆ Continue to lay the vertical bricks, but set the last six bricks dry and equally spaced, then mark their positions on the course below and remove them.

◆ Mortar the six bricks in place.

◆ For a gable wall, raise the wall to the eave level, then erect three-course leads at both ends of the topmost course *(inset),* cutting the outer bricks at an angle to fit snugly under the sloping eave.

◆ String a mason's line between the leads, anchored into the mortar with nails, then lay the intermediate bricks, anchoring the bricks to the wall with a wall tie every third course; repeat the process until the veneer wall is completed.

◆ Cover the ragged edge of the sloping brickwork with a strip of molding or an overhanging frieze board.

# A ROWLOCK SILL

**1. Setting the end bricks.**

◆ Seal the $4\frac{1}{2}$-inch opening below each window with a 12-inch-wide strip of 6-mil polyethylene sheeting, nailed to the backup wall *(page 83, Step 2).*

◆ Cut two bricks to 5 inches in length and mortar them in place at the ends of the opening, angling them slightly downward. Check that both bricks protrude equally from the wall face—usually $\frac{1}{2}$ inch—and hold a level across the tops of the bricks to be sure they lie at the same angle.

**2. Adding intermediate bricks.**

◆ Cut more 5-inch bricks and set them in place dry to space the joints, marking their positions on the window sill.

◆ Adjust the mason's line until it runs level with the tops of the end bricks, holding it flush by folding scraps of cardboard around the line at the end bricks and clamping them down under scrap bricks placed on top of the end bricks.

◆ Butter and lay the sill bricks, aligning them with the marks and the line *(above)*, and inserting pieces of rope or $\frac{3}{8}$-inch tubing at every second mortar joint *(page 83, Step 2)*.

◆ Finish the mortar joints *(page 76, top)*.

## A BLOCK WALL FACED WITH BRICK

When brick veneer is backed by a concrete-block wall, the bricks and blocks rise together, with a 1-inch air space between them; the local building code will dictate the total required thickness of the wall. Both the concrete-block backup wall—here 4 inches thick—and the brick veneer share a concrete footing *(pages 44-50)*. The two layers are tied together at every sixth brick course with joint-reinforcement wire straddling both layers. Because no existing wall is present to support guide posts for the courses, stepped-back corner leads are erected for both layers. Steel joint-reinforcement wire is installed in the leads, cut 12 inches longer than the lead so that it will overlap the reinforcement wire in the intermediate blocks to be laid later. Flashing is also installed in the appropriate courses of the corner leads, and is cut 4 inches longer than the lead to overlap the flashing on the bricks between the leads. The flashing is placed in the same locations as for brick veneer over wood siding, and is anchored to mortar joints in the backup wall.

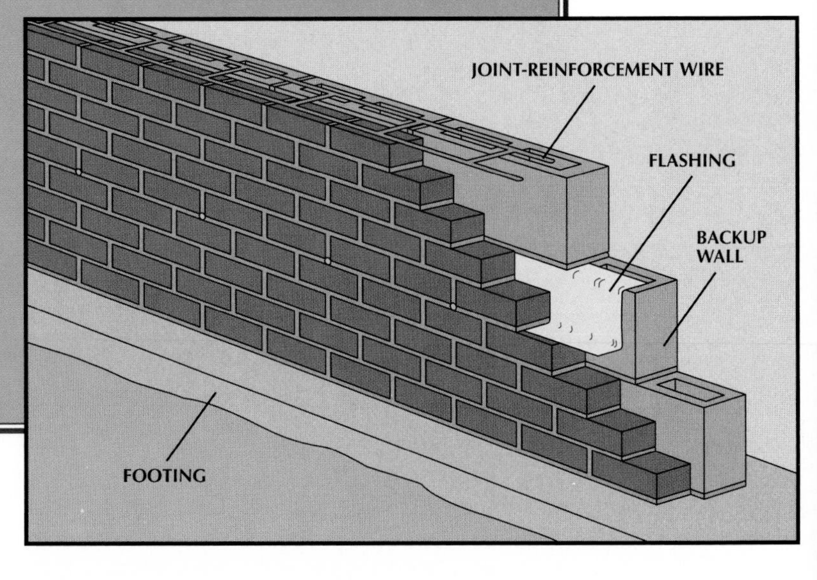

# Building a Stone Wall

Much of stonework's beauty comes from the random mosaic created by the uniqueness of each stone's shape. This irregularity of shape, however, means that building is slow-going, since providing firm support for each piece as it is being laid is essential.

The bottom course in any stone wall—whether freestanding *(below)* or a veneer *(page 90)*—must be supported by a sturdy frostproof footing *(pages 44-46)*. The stones lying on the footing are bedded solidly in mortar to prevent shifting or slipping. Standard Type N mortar is best, although a stiffer mixture is required than for bricks or blocks, because of the stones' weight.

**Sorting and Laying Stones:** Save as corner pieces stones with two flat faces that meet at a neat 90-degree angle or any that can be easily squared by cutting *(page 14)*. Make a mental note of the proportion of roughly rectangular stones to those without any particular shape, and the proportion of large stones to small ones. As you raise the wall, incorporate different sizes and shapes for a balanced appearance.

When laying stone, proceed slowly, evaluating each piece for the aesthetic and structural opportunities it presents. Trowel a firm, flat bed of mortar for each piece to rest on. Position stones on the wall so that their top surfaces are level or slope down toward the core.

The first course is kept in alignment by sighting along building lines strung from batter boards *(pages 34-38)*; for further courses a mason's line is stretched between line pins pegged in the mortar joints. As you work, stagger vertical joints wherever possible, and always try to place a large, heavy stone over two or three smaller ones to spread and balance its weight.

**TOOLS**

- Maul
- Hammer
- Handsaw
- Concrete tools
- Mason's trowel
- Level
- Tuck pointer
- Line pins
- Mason's line
- Mason's hammer
- Cold chisel
- Pointing trowel
- Stiff-bristle brush

**MATERIALS**

- 1 x 6s
- 2 x 4s
- Stakes
- Form boards
- Common nails (2")
- Concrete
- Mortar
- Stones

**SAFETY TIPS**

*Wear gloves while working with mortar and hard-toed shoes when handling stone.*

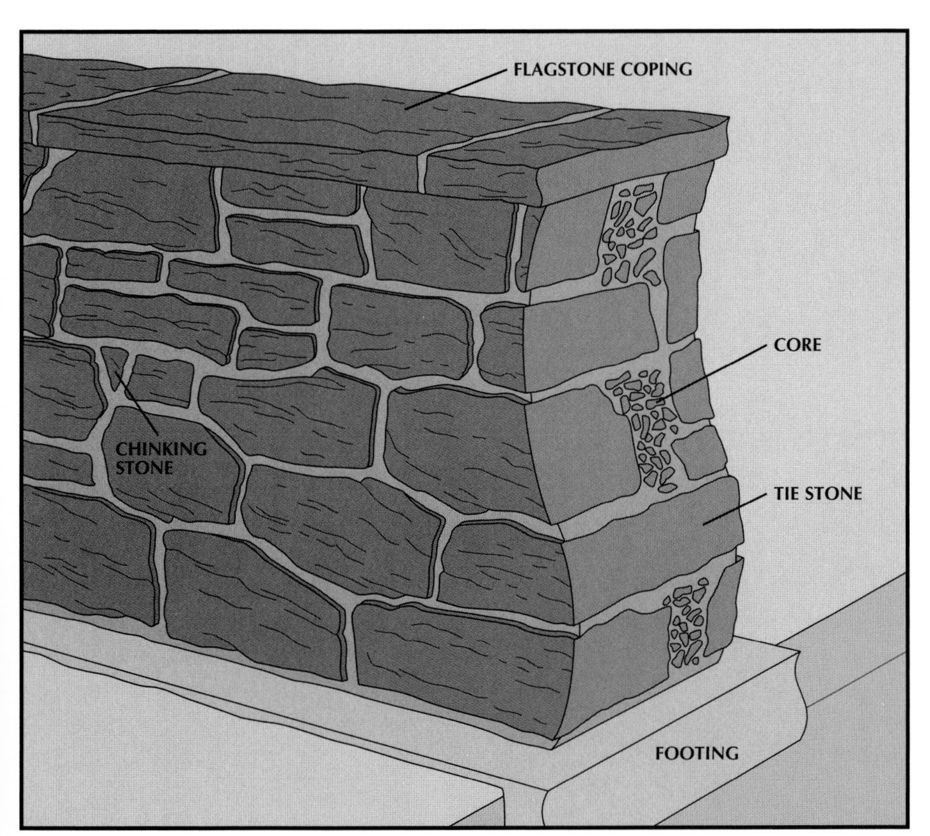


**Anatomy of a solid stone wall.** A stone garden wall consists of two faces of large and small rocks sitting squarely upon a wide, poured-concrete footing *(pages 44-50)*. A thick, level bed of mortar joins the footing and the first course—large stone blocks that fit together easily. The remaining courses comprise large and small stones mixed together. Irregular stones—called chinking stones—fill large gaps. Stone chips and mortar fill the core between the two faces of the wall, as seen in the cutaway view at left. Long tie stones cross over the core every few feet to bond the two wall faces together. For extra strength, vertical mortar joints are staggered from course to course. A $2\frac{1}{2}$-inch-thick flagstone coping, mortared in place, provides a weatherproof cap.

## LAYING THE COURSES

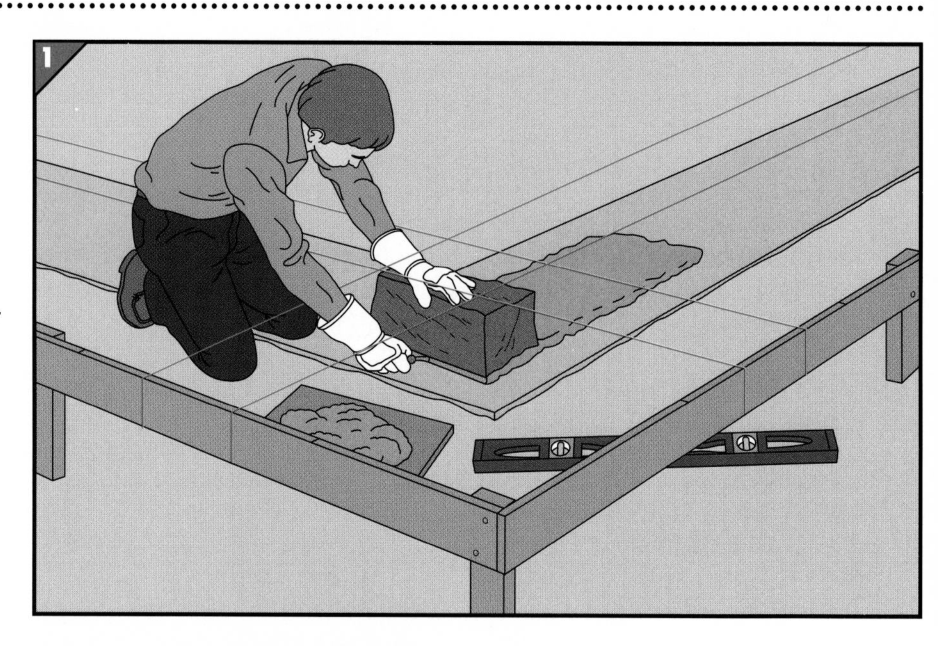


**1. Setting the cornerstones.**

◆ Stretch strings between the inner and outer building lines on the batter boards *(page 37)*.
◆ Spread a 1-inch-thick mortar bed on the footing, aligning it with the strings.
◆ Set a cornerstone on the mortar, aligning it with the outer building lines.
◆ With a tuck pointer, force more mortar beneath the stone until the two corner faces of the stone are plumb *(right)*.
◆ Set the other corners the same way.
◆ Lay the first course of the inner and outer walls with large stones that fit together easily, spacing them $\frac{3}{4}$ inch apart. If you decide to move a set stone, lift it out and wash it thoroughly.
◆ Remove the building-line strings.

**2. Filling vertical joints.**

◆ Scoop up mortar on a trowel and, with the trowel serving as a palette, pack mortar into the vertical gaps between stones with a tuck pointer *(left)*.
◆ Continue filling the joints until the stones are firmly wedged together, but to avoid forcing the stones out of position do not push in too much mortar. If mortar begins to ooze out of a joint, catch it with the point of the trowel and dump it into the core of the wall.

**3. Building up the core.**

◆ Drop small stones and shards into the core of the first course.
◆ Fill the core to the top of the course with mortar, thinned enough so that it is easy to work around the stones *(right)*.
◆ Trowel a new mortar bed on the first course, without smoothing it, and lay cornerstones for the next course.
◆ At the corners of the first course, set line pins into the mortar joints and string a mason's line to keep the face of the wall plumb as you work, then finish laying the course.
◆ Lay each successive course in the same way, moving up the mason's line each time, but in the second course and every 2 feet above it, set tie stones that span both wall faces every 2 feet along the length of the course.
◆ Tool the joints periodically and cap the top of the wall *(page 90, top)*.

# BALANCING STONES

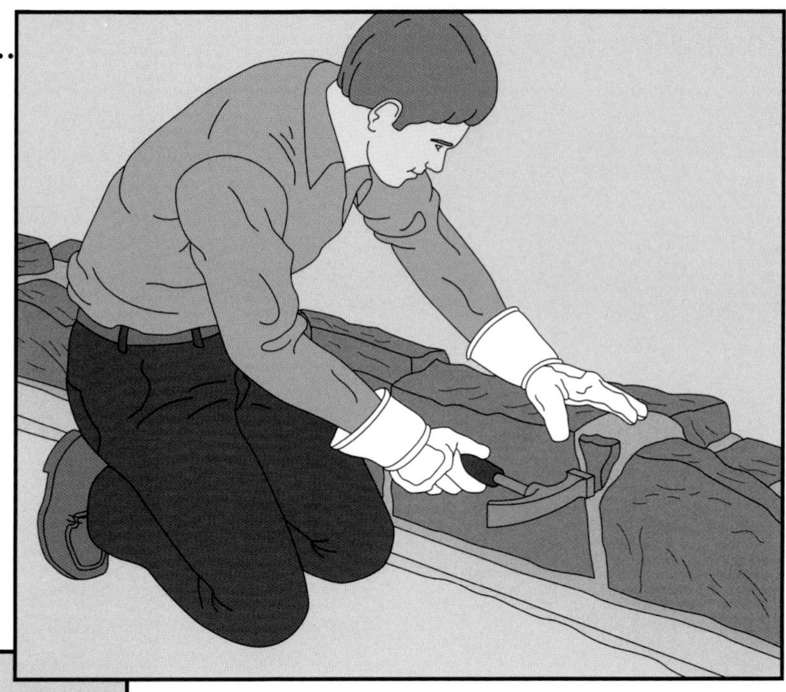

**Chinking large gaps.**
If a large space occurs between two stones, apply a little mortar to the back of a small stone—known as a chinking stone—and tap it into the unsightly joint with a mason's hammer *(right)*. Choose a chinking stone that not only improves the wall's appearance but helps level the top of the course to receive the stone above.

**Shimming stones.**
◆ If the irregular bottom of a stone makes it wobble, wedge a shim—a V-shaped stone chip—into the mortar underneath *(left)*; leave the shim temporarily in place to steady the stone and keep it aligned.
◆ Pack mortar around the shim and, once it has set, knock out the shim with a cold chisel and fill the hole with fresh mortar.

**Shoring stones.**
◆ For a teetering stone that is too large to be steadied with a stone shim, position a length of 2-by-4 at the correct angle to support the stone, then anchor the board at ground level with a heavy rock *(right)*.
◆ Pack mortar in the space under the loose stone.
◆ When the mortar has set, remove the wooden brace.

**Tooling joints and adding a cap.**

◆ After the mortar in each joint has been allowed to set for 30 to 45 minutes, finish the joints with a pointing trowel, adjusting the depth of each joint to the size and shape of the stones that surround it.

◆ With a stiff-bristle brush, such as an old paintbrush or a whisk broom, smooth the remaining mortar *(left)*.

◆ Cover the top by spreading a bed of mortar and laying $2\frac{1}{2}$-inch slabs of flagstone that span both wall faces.

## A SURFACE STONE TREATMENT

A single thickness of decorative stone rests on the same footing as the concrete-block wall that it covers. The pattern of the stones reflects the masonry principles used to erect a freestanding stone wall—although a veneer wall does not support a structure, it must support itself. Stones are set firmly in position, with mortar at their backs as well as above and below. Wall ties spaced as for brick veneer *(page 84, Step 3)* strengthen the bond between the block wall and the stone veneer. They are set in the wall's mortar joints as it is being built. As the stone veneer is raised, you will need to figure in advance the size and placement of stones around the protruding wall ties, so that the ties end up positioned as close to mortar joints as possible.

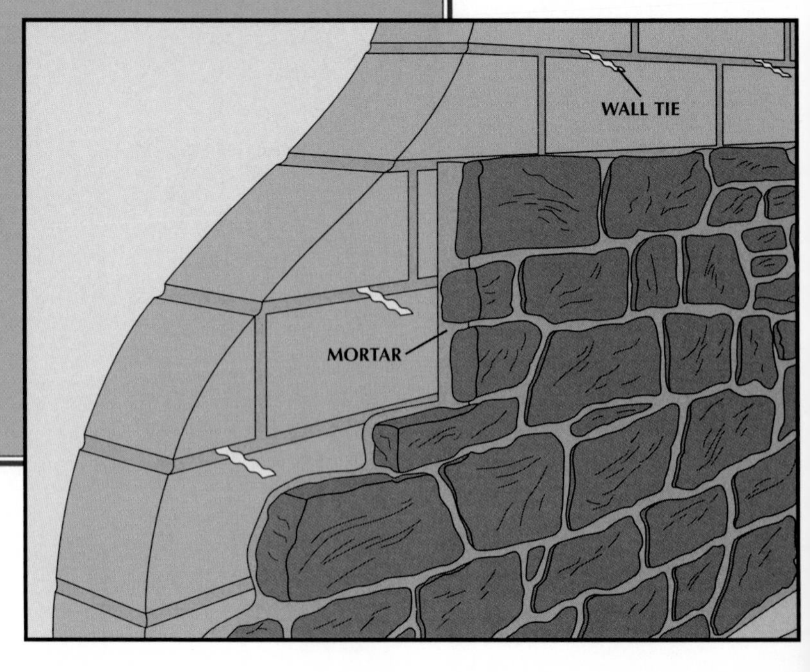

# A Block Wall to Retain Soil

**H**olding back a mass of earth—whether to control erosion or finish the edge of a lawn or driveway—calls for a sturdy retaining wall. These walls can be built with most of the same techniques as ordinary freestanding concrete-block walls *(pages 70-80)*, but the stresses placed on them demand a few adaptations.

**Planning:** Designing a wall to withstand the enormous pressure of rain-soaked earth requires expert knowledge, so plans for any retaining wall higher than 4 feet must be drawn up by a structural engineer. Even for a lower wall, you may want professional advice. Consult a building engineer, or use one of the preapproved designs often furnished by the local building department.

The simplest styles to build are cantilever and gravity walls *(page 92)*. A cantilever wall, which requires vertical reinforcement, is so named because it is structurally tied to one end of a footing that extends far back into the earth. A gravity wall relies on its great mass of masonry for strength. Because more material and labor is involved for a gravity wall than for a cantilever wall, they are seldom built more than 3 or 4 feet high. If either of these walls will be subjected to extreme pressure, they may need abutments; such projects are best left to a professional.

**Water Drainage:** Gravity walls have a sloping perforated pipe behind them that carries water away from the wall *(page 93)*. In cantilever designs a pipe can be used or water can be allowed to escape the soil behind the wall through weep holes—small openings in the wall *(page 93)*. The top of either wall is capped and the back parged, or coated, with damp-proofing compound to divert seepage to the weep holes or drainpipes *(page 94)*.

**Special Building Techniques:** When laying the blocks, mortar the cross webs as well as the face shells of the blocks with type S or M mortar. Allow the mortar to cure for at least 10 days before backfilling the area behind the wall. Usually the backfill is the soil that was removed during the wall's construction, but in some cases the engineer may specify a free-draining material such as gravel or granular soil. Coarse gravel or stone is laid behind the weep holes and under the drainpipe to prevent them from becoming clogged with dirt.

**TOOLS**

Mason's trowel
Mason's line
Shovel
Hacksaw
Tin snips
Leaf rake
Paintbrush

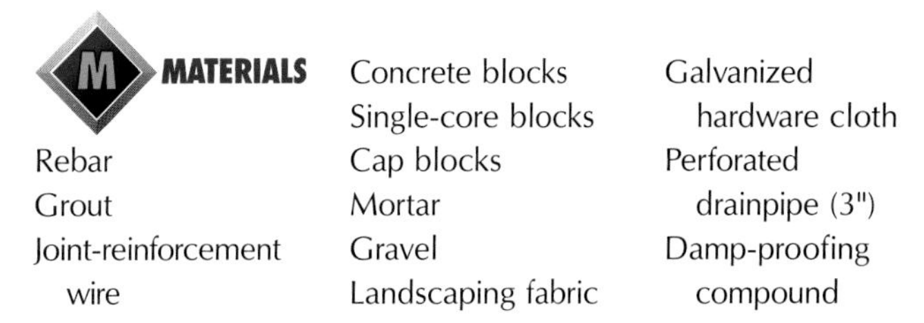

**MATERIALS**

Rebar
Grout
Joint-reinforcement wire
Concrete blocks
Single-core blocks
Cap blocks
Mortar
Gravel
Landscaping fabric
Galvanized hardware cloth
Perforated drainpipe (3")
Damp-proofing compound

**SAFETY TIPS**

*Put on gloves when working with mortar, and wear hard-toed shoes to protect your feet from dropped blocks.*

## A MODULAR-BLOCK RETAINING WALL

Special modular concrete blocks are ideal for retaining walls. The wall rests on a base of sand, and the rows of blocks are built up with each one slightly offset so the wall effectively "leans" against the soil. For the type of blocks shown below, each one is linked to the block above with matching grooves and ledges. Furthermore, the blocks are tapered, making it possible to create curved walls. Depending on the type of block used, these walls can be built to a height of 24 to 48 inches without anchoring. Taller walls need special mesh material placed behind them to hold back the soil.

Because the blocks are not mortared together, water can filter between them, and they do not require weep holes. In areas with poorly draining soil, water accumulation behind the wall can be prevented by setting a drainpipe in the gravel at the level of the first course of blocks.

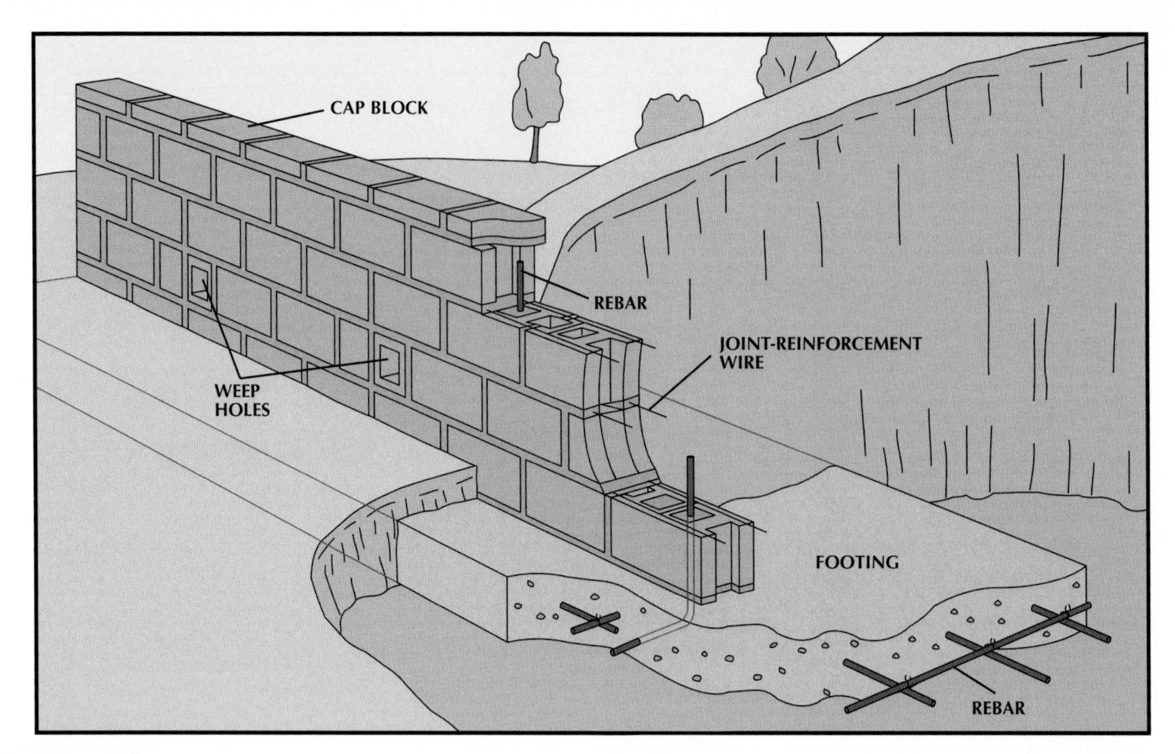


**A cantilever wall.**
This type of retaining wall is placed near the front of a reinforced footing *(page 47)*, and is secured to it with vertical reinforcing bars cast into the footing: One end of each vertical rod lies horizontally within the footing, as described on page 48, and the other rises through the bottom two courses of concrete block. Additional lengths of rebar, lowered through the hollow blocks, lock together the courses. The cores of any blocks through which the rebar passes are solidly filled with grout. Near the bottom, a row of weep holes allows water to drain from behind the wall *(opposite)*. Strips of joint-reinforcement wire buried in each horizontal mortar joint help prevent cracking *(page 74)*. Some designs require a bond-beam course to stabilize the top of the wall *(pages 79-80)*. For further waterproofing, the wall is capped with solid blocks and the back face is parged *(page 94)*.

**A gravity wall.**
This wall rises in steps, with each course one block narrower than the one before. The structure is mortared to its reinforced footing *(page 47)* and the faces of the blocks are mortared to each other. A slanting drainpipe embedded in gravel carries away water that collects behind the wall *(opposite)*. The hollow top blocks of each face are capped with solid blocks to seal out water, and before backfilling, the steps on the back surface of the wall are parged *(page 94)*.

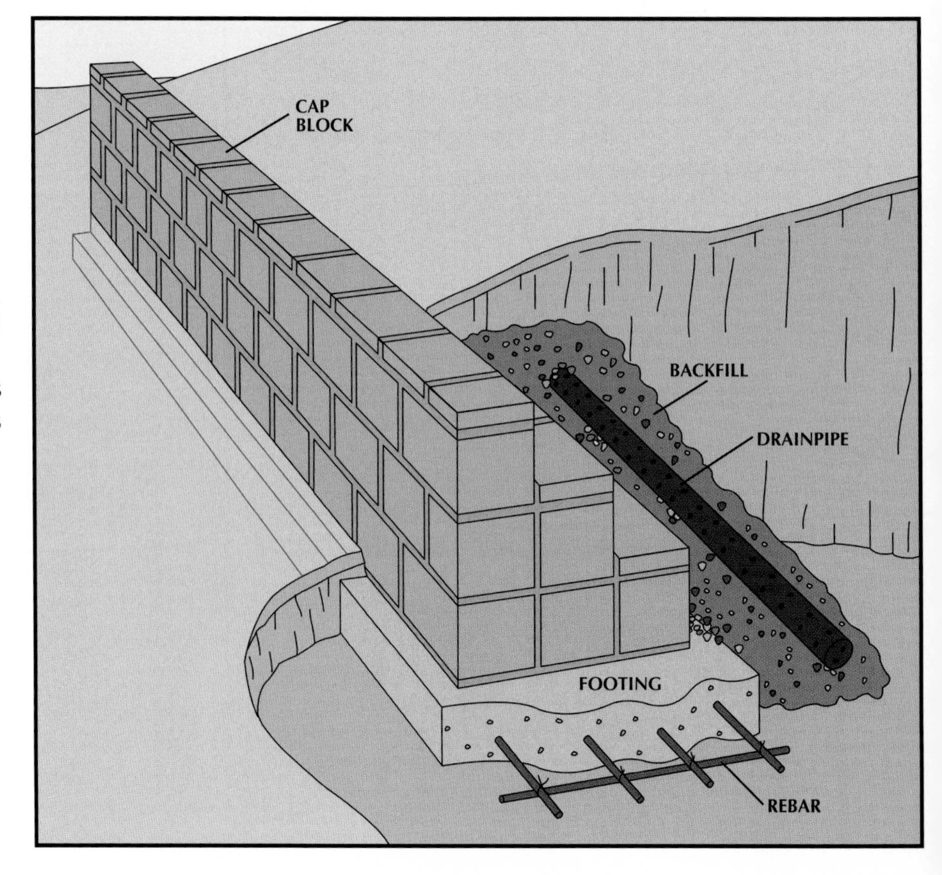

**Building in weep holes.**

◆ At every point where the plan calls for a weep hole, lay a square, single-core block on its side so that the core opens through the wall *(above)*. Next to this one, lay a second square block, core up, to reestablish the running-bond pattern of the wall. Lay a full course of ordinary blocks on top of the weep-hole course.

◆ Trowel mortar around the back edges of the single-core blocks and embed a square of coarse galvanized hardware cloth or screening in the mortar to keep debris out of the weep holes.

When the wall is complete *(page 94)* and the mortar has cured, backfill with soil until you reach the top of the course just below the weep holes. Then spread gravel behind the entire course containing the weep holes; extend the layer of gravel to the back edge of the excavation *(inset)*. Finish backfilling with dirt above the gravel. With very fine soil, a layer of landscape fabric between the soil and gravel can keep earth from filtering into the gravel.

**Laying a drainpipe.**

◆ When the wall is finished *(page 94)* and the mortar has cured, backfill behind the wall to a level about 12 inches above the footing, mounding the soil about 6 inches higher at the center of the wall.

◆ Add 3 inches of gravel or crushed stone.

◆ With a hacksaw, cut two lengths of 3-inch perforated pipe long enough to reach from the top of the mound to daylight 6 inches beyond the end of the wall.

◆ Place the pipes on top of the gravel on each side of the mound *(above)*, and bend a square of galvanized hardware cloth or screening over the high end of each pipe.

◆ Cover the pipes with 3 inches of gravel or crushed stone.

◆ Finish backfilling with soil.

**Capping the wall.**

◆ After laying the last course of hollow blocks, lay one solid cap block, 2 to 4 inches thick, atop each end of the wall.
◆ Measure the distance between these two blocks, and mark the center point between them with chalk.
◆ Stretch a mason's line between the two cap blocks and, with the line as a guide, lay cap blocks from one end of the wall toward the center point *(right)*. Stop when the last block is at the center point.
◆ Lay blocks from the other end of the wall to finish the cap.

**Parging the back of the wall.**

◆ Dampen the exposed horizontal part of the footing, as well as the back of the wall, to the eventual ground level.
◆ Trowel on mortar at the joint between the wall and the footing, shaping the mortar into a rounded cove.
◆ Trowel a layer of mortar about $\frac{1}{4}$ inch thick over the entire back surface of the wall, stopping about 6 inches below ground level.
◆ Roughen the mortar with a leaf rake or a trowel; after 24 hours, add a second coat of the same thickness.
◆ After the wall has cured for about 3 days, use a paintbrush, roller, or trowel to apply a coat of asphalt-base damp-proofing compound over the mortar *(left)*.
◆ After the time specified in the directions for the compound, apply a second coat.

# Decorative Brick

**A**bove all else, a brick wall is functional, but it can be made decorative by altering the traditional bonds, or patterns, in which bricks are laid, or by combining bricks of different colors.

**A Variety of Looks:** The simplest way to achieve a decorative appearance is to cap a wall with specially cast coping bricks *(page 9)*, or to tint the mortar. Also, the mortar joints can be finished with different tools to vary their shape, or left unfinished *(page 98)*. These unfinished "weeping joints," however, will not withstand harsh climates.

Bricks can be offset to create an interplay of light and shadow, or to form a quoin—a design that accents a corner *(pages 100-102)*. One decorative variant of brickwork is to make a lattice *(pages 103-106)*; another is to create graceful serpentine curves *(pages 107-111)*. Lattice and serpentine walls are laid in a single-brick thickness, and are generally no taller than 4 feet because of the consequent loss in strength. A serpentine wall in effect braces itself, but a lattice wall is buttressed every 6 to 10 feet with brick piers 12 to 16 inches square. For single-thickness walls, choose bricks with two finished faces. Tie the inner and outer walls of a double-thickness structure together with wall ties laid in the mortar every 12 inches along every second course.

**Planning Ahead:** Any decorative brick wall requires a poured-concrete footing *(pages 44-50)*. Check that your planned structure conforms to local building codes, and if the wall will rise near a property line, discuss your plans with your neighbor. To help with the ordering and setting of bricks, make a preliminary sketch on graph paper of one complete unit of the design, considering the variety of options for bricks *(pages 8-9)*.

**TOOLS**

- Mason's trowel
- Grapevine jointer
- Mason's ruler
- Tuck pointer
- Mason's level
- Story pole
- Mason's line
- Line pins
- Chalk line

**MATERIALS**

- Mortar
- Bricks
- Wall ties

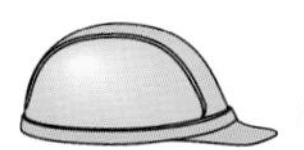

**SAFETY TIPS**

*Protect your hands with gloves when working with mortar. Wear steel-toed shoes when handling bricks.*

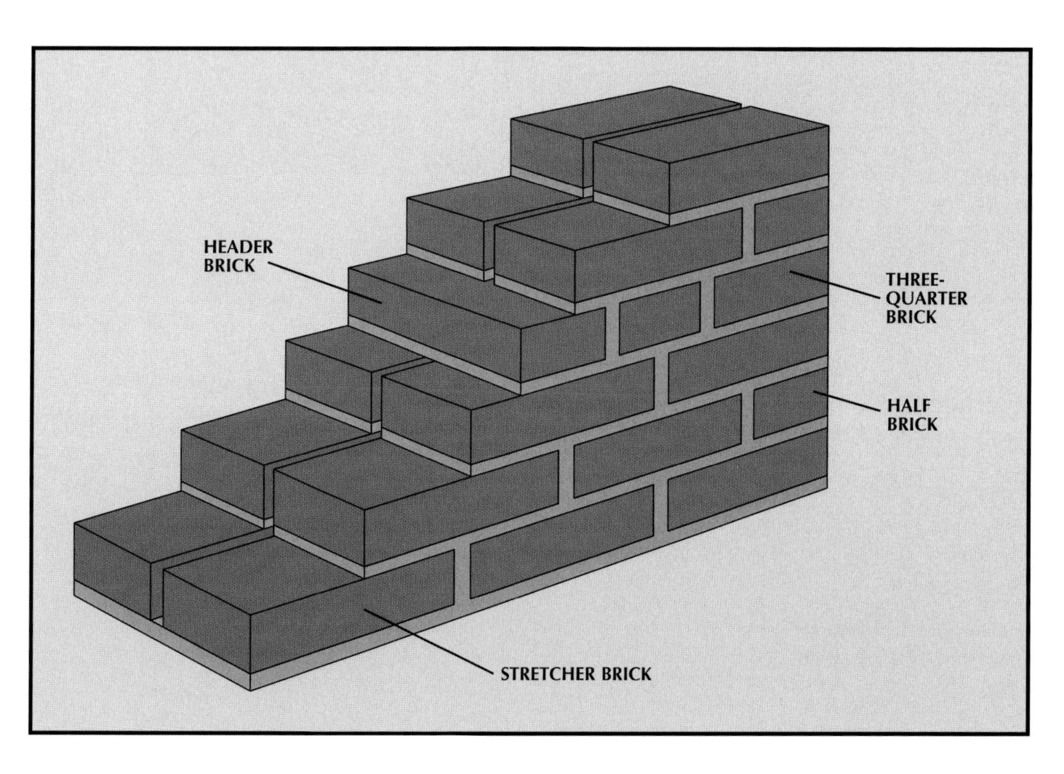


**Bricklaying terms.** The name given to a brick laid in a wall comes from its position. A stretcher brick is placed lengthwise, with its long narrow side forming the face of the wall. A brick laid crosswise with its ends facing outward is called a header brick. Courses of headers strengthen a double-thickness wall by binding the two layers together; for a single-thickness wall, you can substitute half bricks for the header courses. Half bricks and three-quarter bricks are also used to bring the courses even at the ends of walls.

**Common bond.**
Strong and easy to lay, common bond has a row of headers interrupting conventional rows of stretchers after every fifth course. Whole bricks, half bricks, or three-quarter bricks are laid at the ends of the wall as needed, in order to bring the courses even.

**English bond.**
A decorative pattern that makes an exceptionally sturdy wall, English bond alternates courses of headers and stretchers.

**Flemish bond.**
Header bricks are alternated with stretcher bricks within each course in Flemish bond. As with common and English bonds, the strength of this wall is enhanced when the bricks are laid in this pattern.

**Flemish-cross bond.**
Standard Flemish courses alternate with courses consisting entirely of stretchers. Each of the headers in the Flemish course is color-contrasted with the stretchers in the same course, and aligned with every third brick in the stretchers-only course.

**Flemish-spiral bond.**
This design bands the wall with diagonal lines of contrasting headers. Constructed of standard Flemish courses, the bricks are laid so that the headers in successive courses are staggered one half their width beyond the headers in the course below, with every course staggered in the same direction.

**Garden-wall bond.**
Here, Flemish courses are modified, with three stretchers separating headers, to create a large diamond pattern. With each course, the row of contrasting bricks is lengthened by a half brick for five rows; each successive row is shortened by a half brick. For such dovetailing diamond patterns, a preliminary sketch helps you determine the proper placement of the header bricks that form the top, bottom, and center of each of the diamonds.

**A grapevine joint.**

◆ An hour or two after laying the bricks, fit a special ridged grapevine jointer *(photograph)* in the vertical mortar joints of your wall, pressing down firmly and drawing the jointer slowly along the mortar to create a narrow groove down the center.

◆ Tool the horizontal mortar joints in the same way *(left)*.

**A weeping joint.**

◆ Butter each brick with excess mortar.

◆ Press the bricks firmly into place, forcing out the mortar *(right)*, and taking care not to disturb the joints in the earlier courses.

◆ Allow the mortar that has squeezed out of the joints to cure undisturbed.

# LAYING BRICKS IN RELIEF

**1. Mortaring.**

◆ Throw a line of mortar along the course below the one where you will be laying projecting bricks, and pat the mortar flat with the back of your trowel *(above)*.

◆ Remove any mortar that spills over the edge of the course by trimming it away with the edge of the trowel.

◆ Lay the next course, positioning the offset bricks $\frac{5}{8}$ inch—the width of a mason's rule—from the face of the wall, placing the rule beneath the brick as a guide.

◆ Build the wall up in this way, trimming the joints every course or two *(Step 2)*.

**2. Trimming the joints.**

◆ With a tuck pointer or a trowel, carefully trim all the joints around each offset brick flush with the surrounding bricks *(left)*.

◆ Finish all the other joints in the usual way.

# AN OFFSET QUOIN

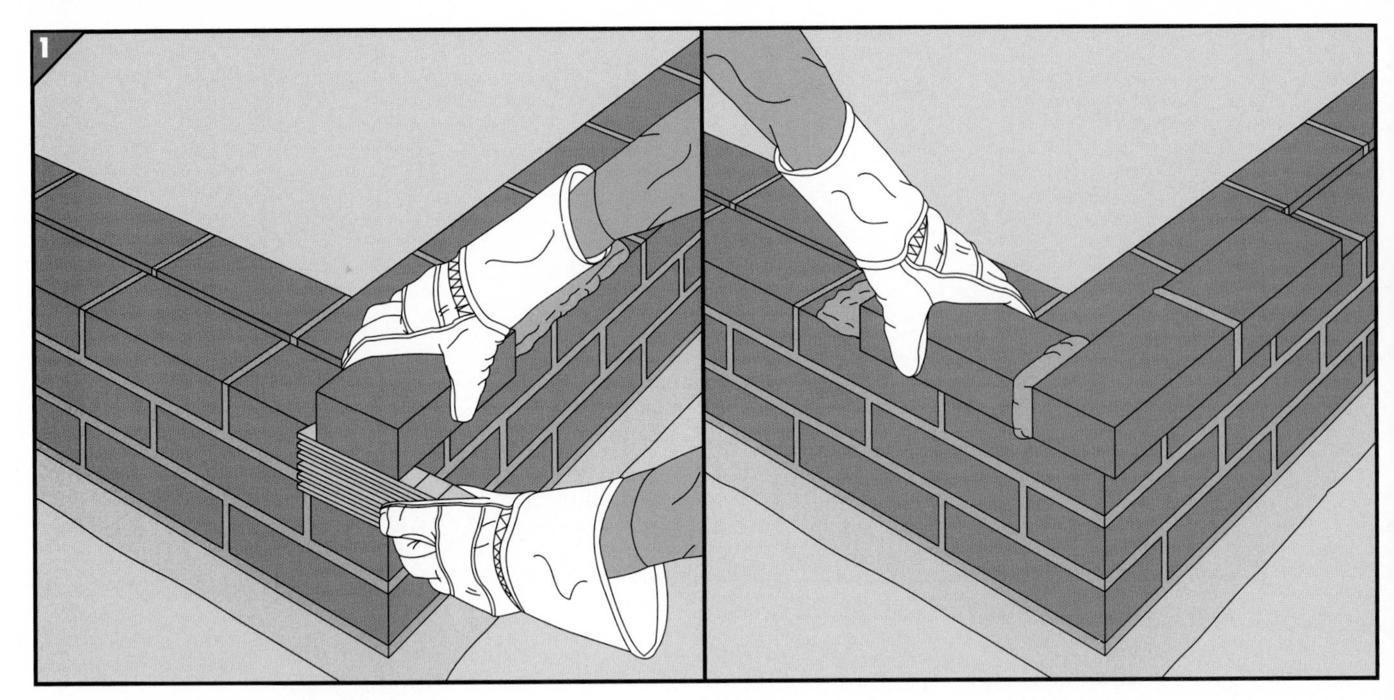

**1. Laying the first offset course.**

◆ Raise both the inner and outer walls to the height of the bottom of the planned quoin.

◆ At the corner, spread a bed of mortar long enough for two bricks along one outer wall—adding wall ties if necessary at this course as you would for a regular wall.

◆ Without furrowing the mortar, lay a corner brick so it is straddling the underlying vertical joint and offset from both faces of the wall by $\frac{5}{8}$ inch—the width of a mason's ruler *(above, left)*.

◆ Lay a second brick end to end with the first; check for horizontal alignment with a mason's level, first along the top of the bricks, then along their outer faces. Adjust the bricks if necessary by tapping them with a trowel handle.

◆ Lay a third brick at the corner, along the other wall *(above, right)*, setting it flush with the end of the first brick and checking its horizontal alignment.

◆ Check the height of the course of bricks using a story pole *(page 73)*.

**2. Completing the first course.**

◆ Along the outer bricks of each wall, spread a mortar bed long enough for three or four bricks and furrow it.

◆ Lay the bricks flush with the face of the wall, making the end joints of these bricks slightly thicker than usual to compensate for the quoin's offset *(right)*.

◆ Check this flush course along both walls with a level and a story pole.

◆ Lay bricks along the inner walls at the corner, flush with the course below.

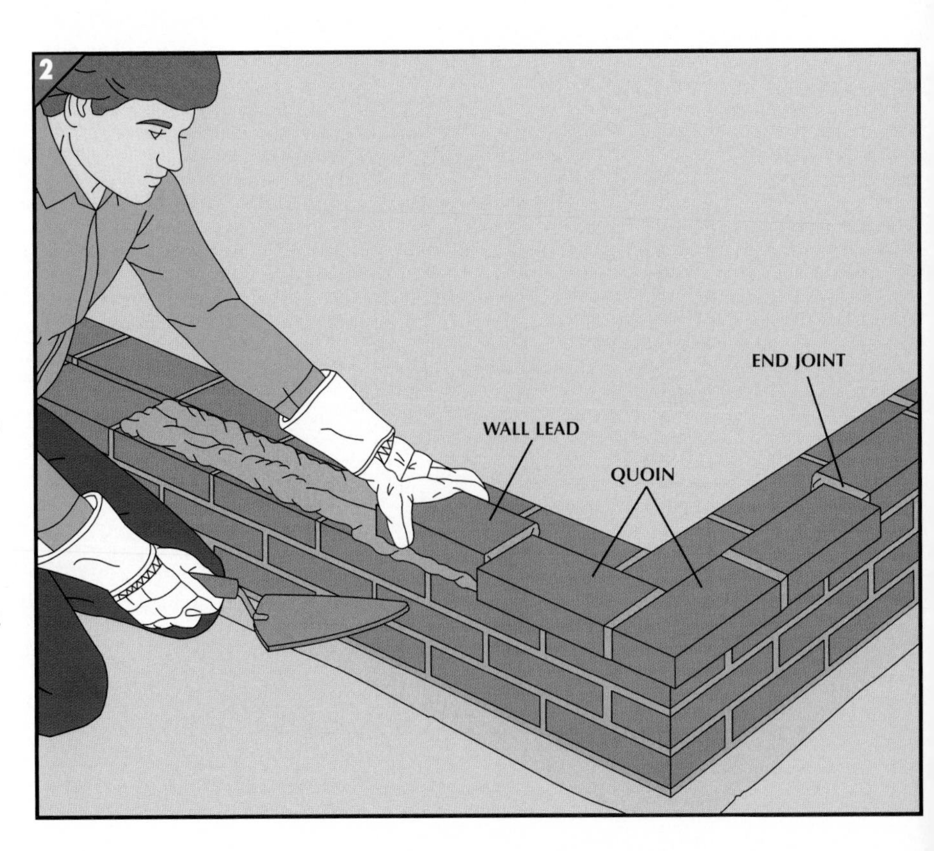

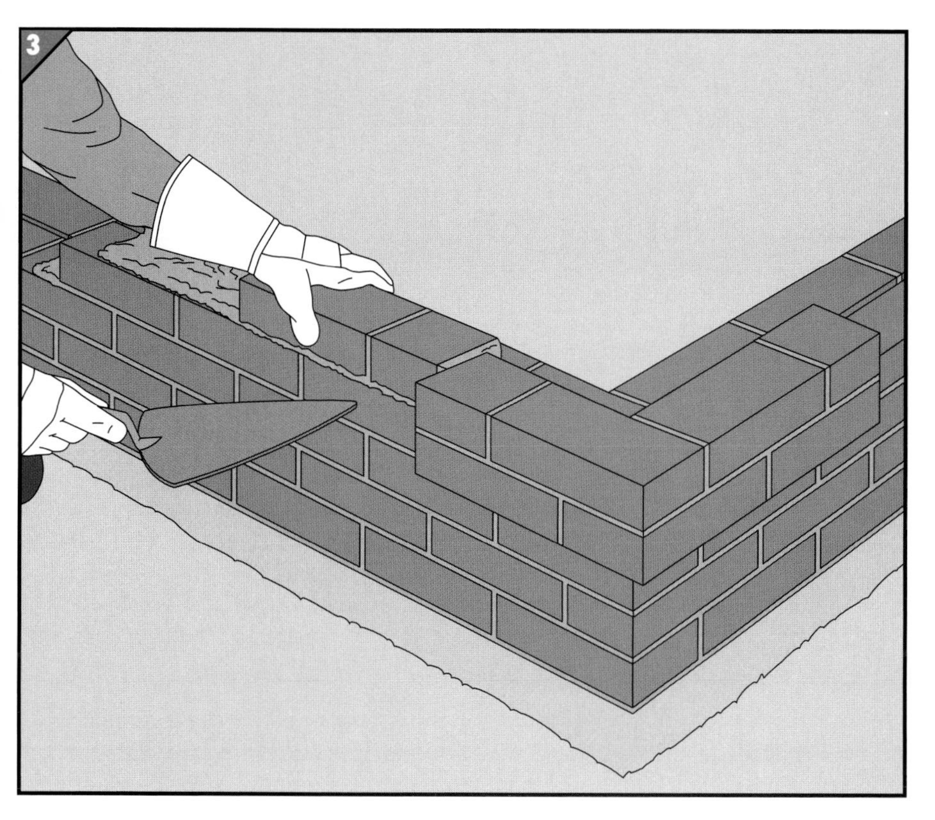

**3. Building up the quoin.**

◆ Lay a second course of offset bricks, positioning a half brick at the end of each leg so that the end bricks of the quoin line up with those underneath.

◆ Build the second course of the lead along both walls, beginning with half bricks and ending a half brick shorter than the first course *(left)*.

◆ Place a second course along the inner wall, setting the bricks flush with the wall surface.

◆ Lay courses of offset bricks until the quoin is the desired height—usually five courses—building up the lead on the outer and inner walls as you go, and setting wall ties at 12-inch intervals at every second course. Check the offset and flush bricks often for plumb and horizontal alignment; align each course with a story pole.

**4. Finishing the quoin.**

◆ At the top of the quoin, lay a course of bricks in nonfurrowed mortar flush with the face of the wall and ending a half brick short of the previous course *(above)*.

◆ Lay a second course of bricks in the same manner, but with furrowed mortar as for an ordinary wall.

◆ Raise up the inner wall to the level of these two courses.

◆ Build a lead, either quoined or conventional, at the opposite end of the wall.

**5. Filling in the wall.**

◆ Run a mason's line along the first course of the leads, anchoring it near the quoins with line pins, and adjusting it to run even with the top of the course.

◆ To compensate for the wide end joint, space bricks evenly in a dry run from lead to lead.

◆ Spread mortar and reposition the bricks *(above)*, then lay a matching course along the inner wall.

◆ Reposition the mason's line along the top of the next course, and continue to raise the wall to the top of the lead, adding wall ties as required.

◆ Fill in holes left by line pins with fresh mortar as you work.

◆ Build up the adjacent wall in the same way.

**6. Adding a second course.**

◆ To start another quoin, lay a course of offset bricks as you did in Step 1.

◆ Following the brick pattern set by these two courses, build up the second quoin and its lead as in Steps 2 through 4.

◆ Fill in the intervening wall as in Step 5.

◆ Continue building the walls in this way, alternating quoins with recessed courses until you reach the desired height *(right)*.

# A MASONRY SCREEN

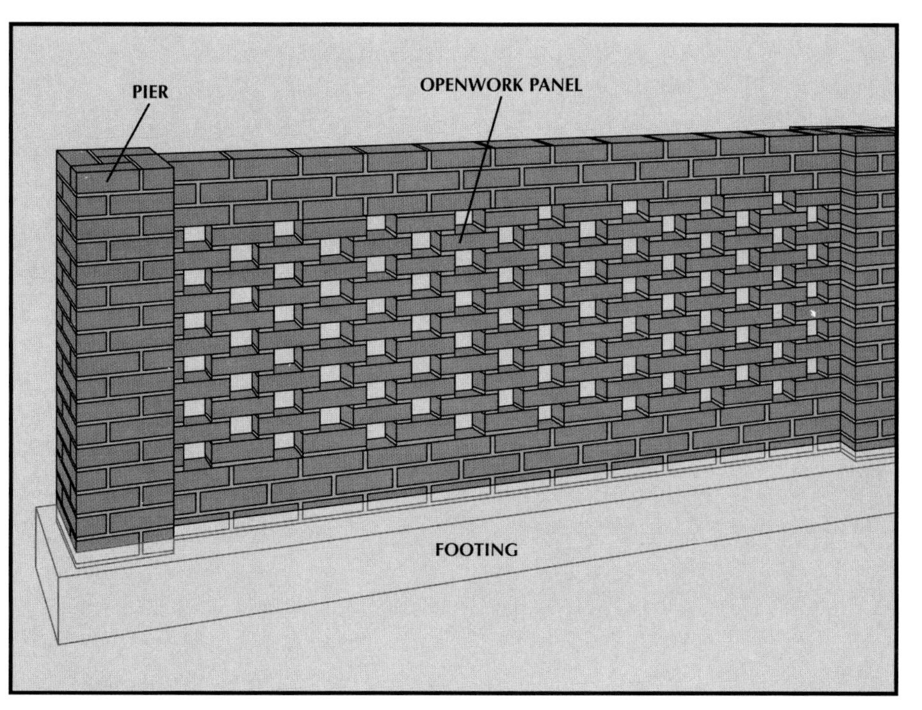


**Anatomy of a latticework wall.** This 18-foot screen, long enough to conceal a carport or to add privacy to a patio, consists of two openwork brick panels, each about $7\frac{1}{2}$ feet long, buttressed by three brick piers, each 1 foot square. The panels, one brick thick, are banded top and bottom by three courses of solid brickwork, to frame the lattice and add stability. With eleven courses of latticework and six courses of solid brickwork, the wall s almost 4 feet high (because of the fragility of latticework masonry, it is best not to exceed this height). Check the local building code for the dimensions of the concrete footing; in this example it is 8 inches thick and 16 inches wide—allowing for 2 inches of clearance on either side of the piers—and it projects 2 inches beyond the ends of the piers.

# ERECTING BRICK LATTICE

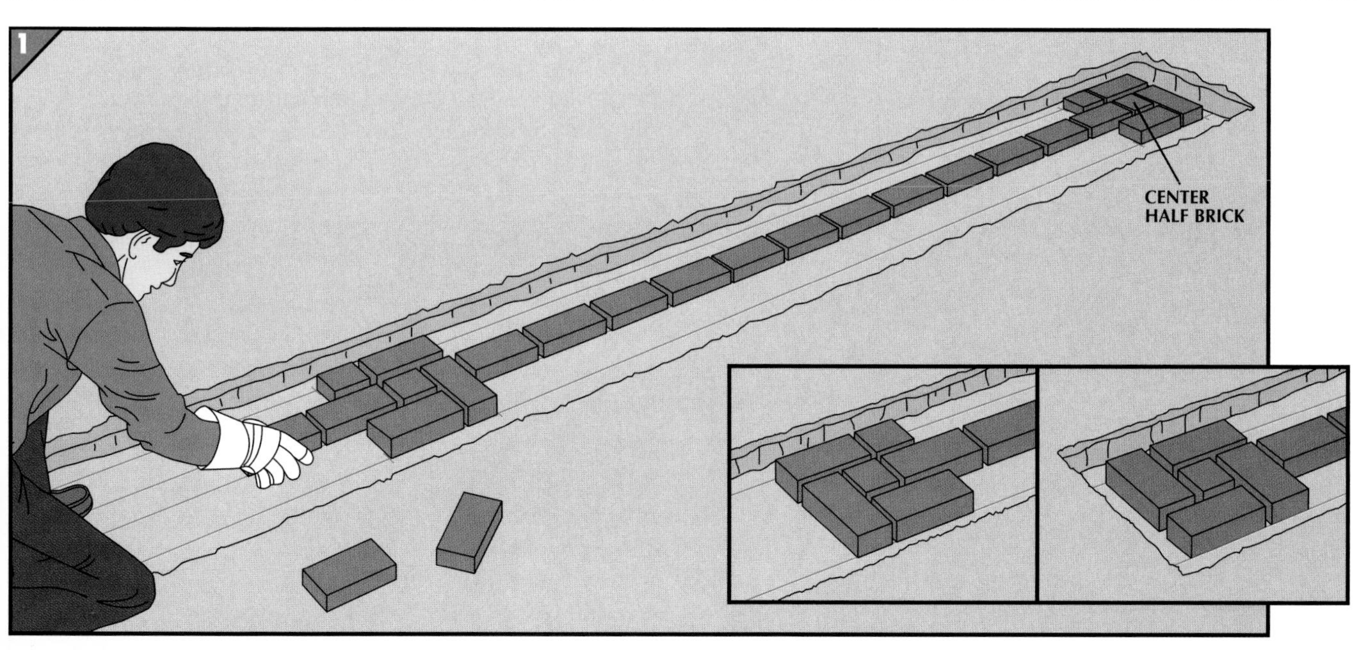


**1. Beginning the latticework.**

◆ Snap two chalk lines along the footing, 12 inches apart, to mark the outer faces of the piers, then snap a second pair of lines, 4 inches in from each pier line and 4 inches apart, to mark the outer faces of the panels.

◆ With a crayon, measure off and mark the points where the piers will cross the panel lines.

◆ Lay bricks in a dry run, beginning with an end pier, pinwheeling the bricks around a center half brick so that the first brick of the wall panel intersects the pier.

◆ Continue placing unmortared bricks between the chalk lines to the middle pier, spacing mortar joints so the last brick falls either a mortar joint away from the pier mark or halfway across it. If the last brick falls a mortar joint away from the pier mark, the pier pattern will follow the one in the illustration above and the right inset; if it falls halfway across the pier mark, it will follow the design in the left inset.

◆ Add a half brick and pinwheel the remaining bricks making up the middle pier.

◆ Repeat this procedure to position the bricks for the second panel and the third pier; the pier's pattern will depend on whether the last panel brick ends outside the pier or within it.

## 2. Laying the solid bands.

◆ Lay the first course of each pier in mortar, as well as the first two lead bricks at each end of one of the panels.

◆ Check the thickness of the mortar bed with a story pole *(page 73)*, and the horizontal alignment of the bricks with a mason's level or a 4-foot-long carpenter's level.

◆ Lay two more courses on the piers and the panel leads, alternating the arrangement of bricks on the pier so the first brick of each panel lead steps back by a half brick from the underlying course *(above)*; check your work frequently for alignment, plumb, and course level.

## 3. Completing the bands.

◆ Stretch a mason's line flush with the top of the first course, setting line pins in the mortar joints to hold it in place.

◆ Fill in the intermediate bricks, removing three or four dry-run bricks at a time and replacing them in mortar *(right)*.

◆ Lay the next two courses, moving up the mason's line as you go and patching the holes left by the line pins with fresh mortar.

◆ Repeat the process to lay the solid bands for the second panel.

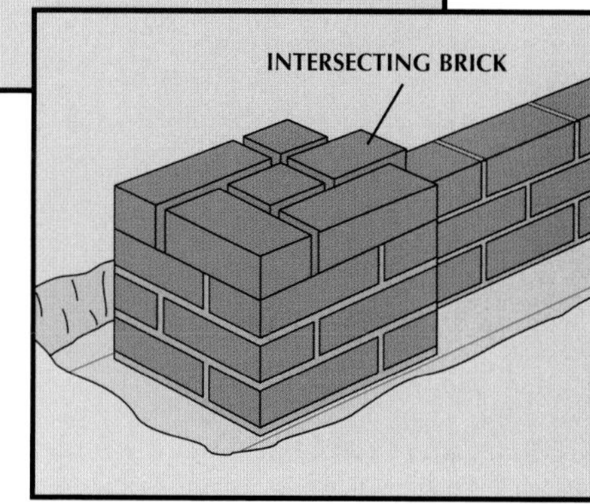


### 4. Beginning the latticework.

◆ At the fourth course, set the pier bricks dry; if the pier is intersected by a panel brick, cut the intersecting brick by 2 inches *(inset).*

◆ Working from two piers toward the center, lay a dry run of panel bricks, about 4 inches apart and spaced evenly *(above).*

◆ Set the pier bricks in mortar, then remove and set the first three or four lead bricks at each end of the panel, laying a mortar bed that is no longer than each of the bricks; remove any excess mortar.

◆ Check the lead for vertical alignment with a story pole and for horizontal alignment with a level.

### 5. Building the leads.

◆ Add a second course of pier bricks, then lay the panel bricks across the 4-inch gaps in the first openwork course; fix them in place with small beds of mortar placed on the ends of the bricks below.

◆ Continue to add bricks in the same way, tying the panels into the piers with bricks shortened by 2 inches every second course and checking frequently for level and for plumb until the openwork leads are five courses high *(left).*

### 6. Filling in the latticework.

◆ Stretch a mason's line flush with the top of the first course of the openwork leads, and set the remaining bricks in the course *(left)*.
◆ Raise the line to match the second course of the leads and fill in the openwork, centering the bricks over the 4-inch gaps in the first course.
◆ Continue in this way until the latticework is even with the tops of the leads.
◆ Raise the second openwork panel to an equal height in the same way.
◆ With a tuck pointer, trim away any mortar that has squeezed into the open spaces *(inset)*.
◆ Continue the latticework upward, building six-course leads, then filling in the openwork; trim away any mortar in the open spaces.
◆ Patch the holes left by the line pins.

### 7. Capping the screen.

◆ Lay a dry run of a new course of solid brickwork along the top of the panels.
◆ Adjust the spacing of the mortar joints so that they are even, then build three-course leads of solid brickwork as in Step 2; take care not to spill the mortar into the open spaces of the latticework as you lay mortar for the first course of the leads.
◆ Check the leads frequently for plumb and level and, with a story pole, make sure the three courses are of uniform thickness.
◆ Fill in the solid brickwork between the leads, with a mason's line as a guide *(right)*.

# A Serpentine Wall

With its unusual complexities of design, a serpentine wall adds grace and charm to its surroundings. Repeated S curves, rather than frequent piers and a double thickness of bricks, give this wall stability.

**Design:** The S curves are a succession of arcs, each an identical segment from a circle of a given radius. The arcs are linked end to end in mirror image down the length of the wall and are terminated by a brick pier at each end of the wall. Since the proportions of the arcs determine the structure's inherent strength, make sure that the radius of the arcs is no greater than twice the height of the wall, and that the total sweep of the arcs from side to side covers a distance equal to at least half the height of the wall.

**Footings:** A concrete footing *(pages 44-50)*, 8 inches thick and resting below the frost line, is adequate. Usually you can dig a simple trench 16 inches wide and 4 inches longer than the wall, but in sandy soil you may need to build curved forms *(page 46)*. For clayey or boggy ground, consult a professional.

**Special Techniques:** Most of the standard bricklaying techniques are employed in building this wall, but since the curves make it impossible for you to use stepped leads and a mason's line to align the courses, you will need to use other methods to check the courses for plumb, level, and proper curve. A plywood template ensures that the curves are identical *(pages 108-109)*. Parallel string lines are set up to mark the outer sweep of the curves, and these strings also serve to plumb and level the courses as you construct the wall *(pages 109-111)*.

**TOOLS**

- Maul
- Saber saw
- Wheelbarrow
- Square shovel
- Float
- Carpenter's square
- Level
- Story pole
- Mason's trowel

**MATERIALS**

- Stakes
- 2 x 2s, 2 x 4
- Plywood ($\frac{1}{2}$")
- String
- Picture wire
- Powdered chalk
- Concrete
- Bricks
- Mortar
- Corrugated-steel wall ties

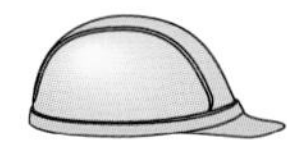

**SAFETY TIPS**

*Put on hard-toed shoes and gloves when handling bricks. Wear long sleeves when applying mortar.*

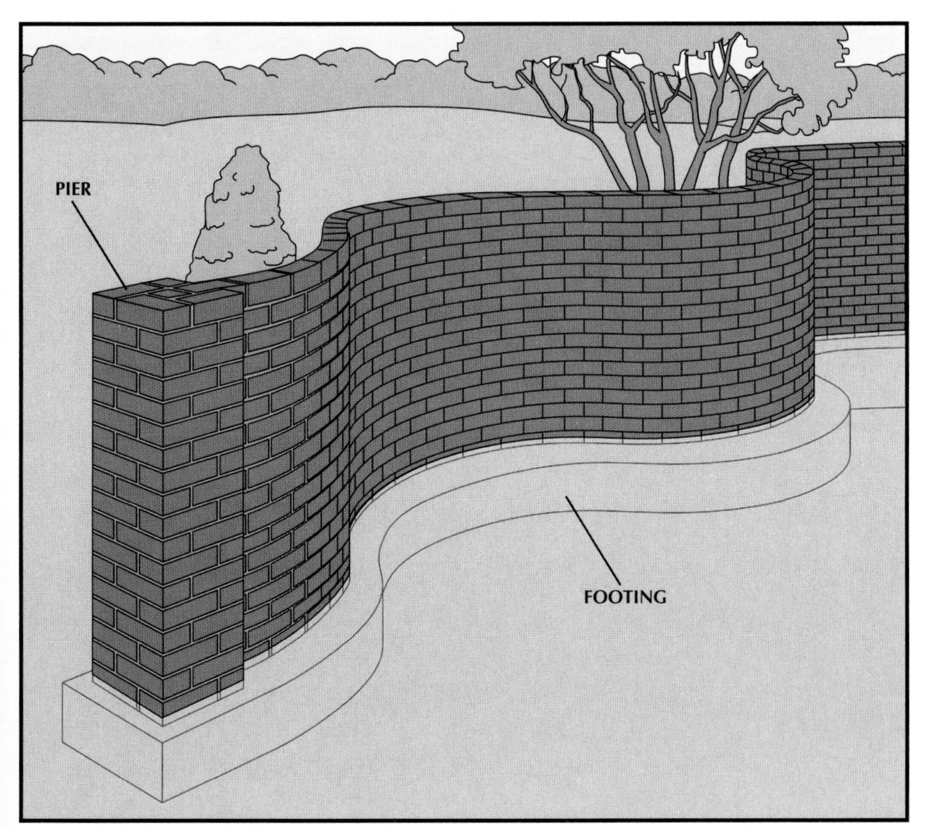


**Anatomy of a serpentine wall.** This traditional serpentine wall consists of one thickness of bricks, built to a height of 4 feet. Each of its undulating curves is a segment of a circle with a radius of 6 feet. The distance between two parallel lines touching the outermost points of the curves is 39$\frac{1}{2}$ inches, which includes the width of the bricks. On each side of the wall, the distance between the outermost points of successive curves is 15 feet 10 inches. Brick piers 12 inches square fortify the ends of the wall, and steel wall ties running from brick to brick along every fifth course provide additional strength. Just below ground level, the wall rests on a similarly curved concrete footing; in this example, it is 16 inches wide, 8 inches thick, and extends 2 inches beyond the wall at each end.

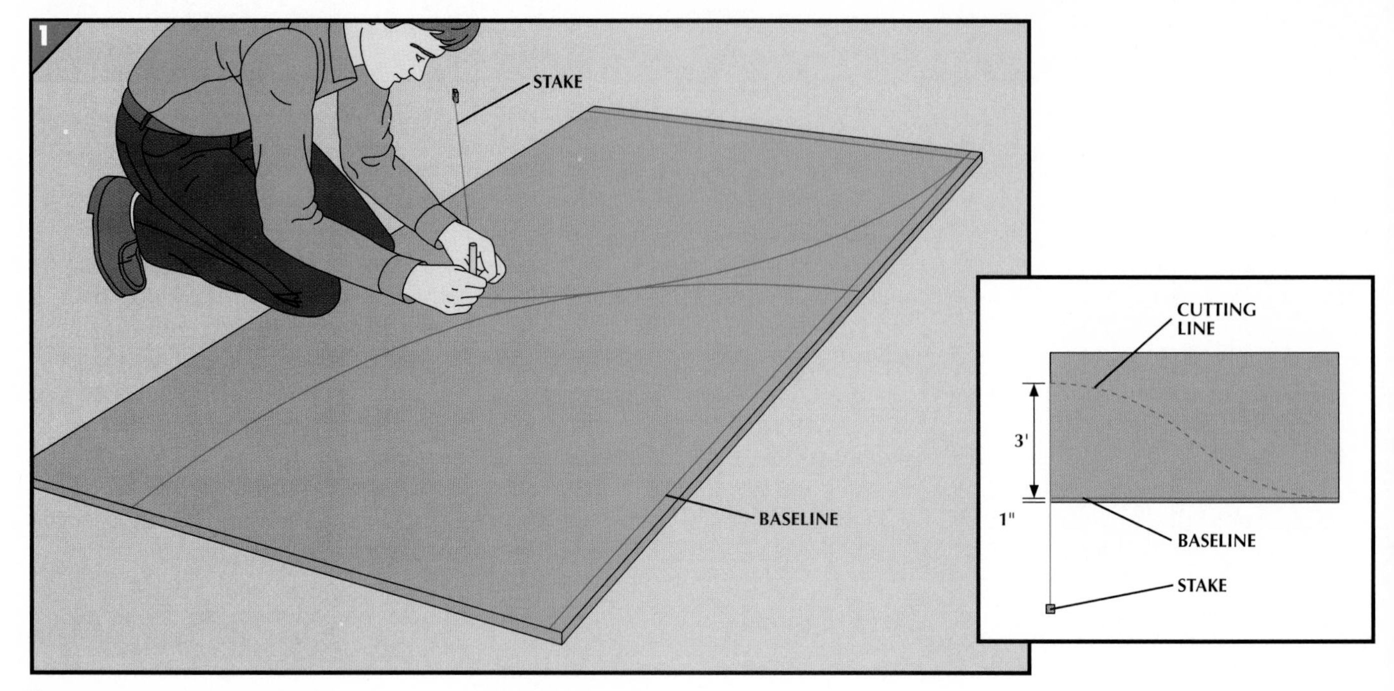


**1. Tracing the template.**

◆ Mark a baseline 1 inch from one of the long edges of a 4- by 8-foot sheet of $\frac{1}{2}$-inch plywood.

◆ Drive a stake into the ground as the center point of a circle and tie one end of a length of picture wire to it. Measure out 6 feet of wire from the stake and twist it around a felt-tipped marker.

◆ Position the plywood so the wire, stretched along one end of the sheet, extends 3 feet past the baseline; with the wire taut, draw an arc across the plywood to mark the first part of the cutting line *(inset)*.

◆ Reverse the plywood and mark a line 1 inch from the end, then adjust the sheet so the wire runs along this line. Beginning at the baseline, draw an arc toward the center of the sheet, extending the curve until the two arcs touch *(above)*.

◆ With a saber saw, cut along both arcs, starting at the end of one arc and switching to the other arc at the point where the two touch. Cut off the 1 inch marked on the far edge where you began the second arc.

**2. Outlining one side of the footing trench.**

◆ Stretch a reference string slightly longer than the planned wall along a line representing the outermost curves of one face of the wall. Secure the ends of the string to two stakes so it lies about 1 inch above the ground.

◆ Lay the template at one end of the planned wall, with the baseline positioned half the width of the planned footing away from the string—in this case, 8 inches.

◆ Outline the curved edge of the template with powdered chalk *(right)*.

◆ Turn the template end over end, mark a baseline 1 inch in from the edge on the other side, and repeat the process to mark off the remaining curves on that side of the planned footing.

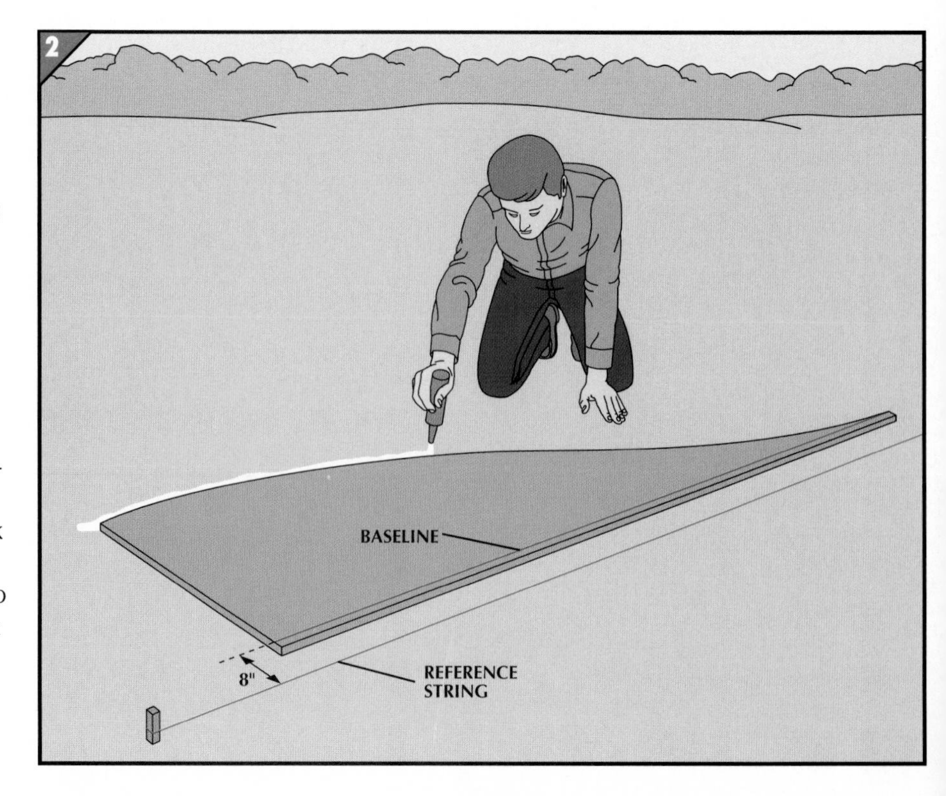

### 3. Completing the outline.

◆ Position the template's baseline half the footing width away from the other side of the reference string and outline a parallel serpentine line *(left)*. The two curved lines represent the full width of the footing.

◆ Extend the curved lines at each end with straight, parallel lines about 14 inches long for the pier footings.

◆ Remove the reference string and dig a footing trench between the lines.

◆ Pour a footing *(pages 44-50)*, and allow it to cure for at least 24 hours before beginning to lay bricks for the wall.

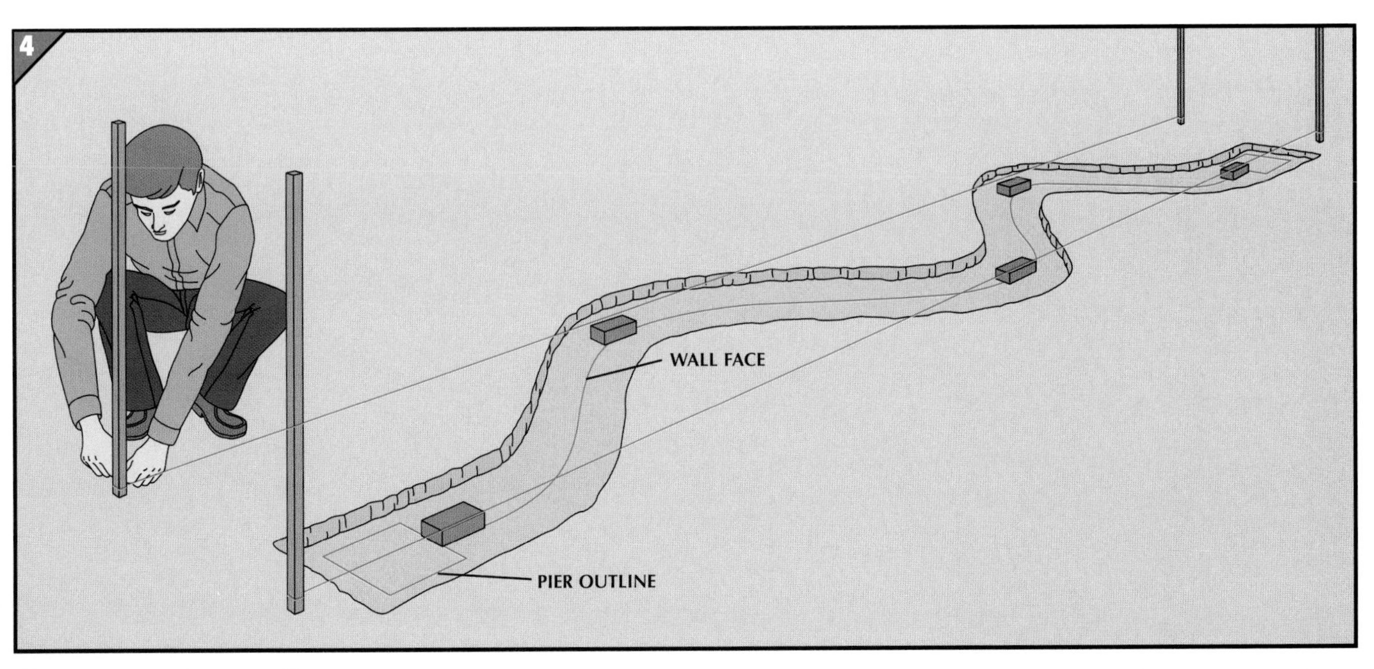


### 4. Setting reference lines.

◆ With a carpenter's square, outline a 12-inch-square pier 2 inches from each end of the footing.

◆ Position the template near one end of the footing so its curved edge is offset from the middle of the footing by the width of a brick. Outline one wall face on the footing, turning the template over as you did in Step 2, and marking the outline with a crayon rather than chalk.

◆ Stretch a string 4 feet longer than the planned wall along the footing so it touches the outermost curves of the crayon line on one face of the wall. Drive a 2-by-2 as tall as the wall into the ground 2 feet past each end of the footing, plumb it, and tie on the string.

◆ Stretch a second string across the footing, parallel to the first and $39\frac{1}{2}$ inches away from it. Place bricks along the crayon line at each end and at the outermost points of the curves to check that the strings are even with the edges of the bricks. Tie the second string to stakes *(above)*, marking the other face of the wall.

# RAISING THE WALL

### 1. Laying a dry run.

◆ With a story pole as a guide *(page 73)*, raise the two strings between the reference poles to a height one course above the footing.
◆ Lay a course of bricks—without mortar—inside one pier outline, pinwheeling the bricks around a center half brick *(page 103, Step 1)*.
◆ Position a dry run of bricks down the middle of the footing, aligning the bricks with the crayon line and setting them $\frac{3}{8}$ inch apart to allow for mortar joints *(right)*.
◆ Adjust the spacing between the last few bricks so the final unit either falls a mortar joint short of the pier outline or crosses the outline at its midpoint, as for a brick lattice *(page 103, Step 1)*.
◆ Lay a dry course of bricks for the second pier, cutting a brick if necessary to interlock with the wall.

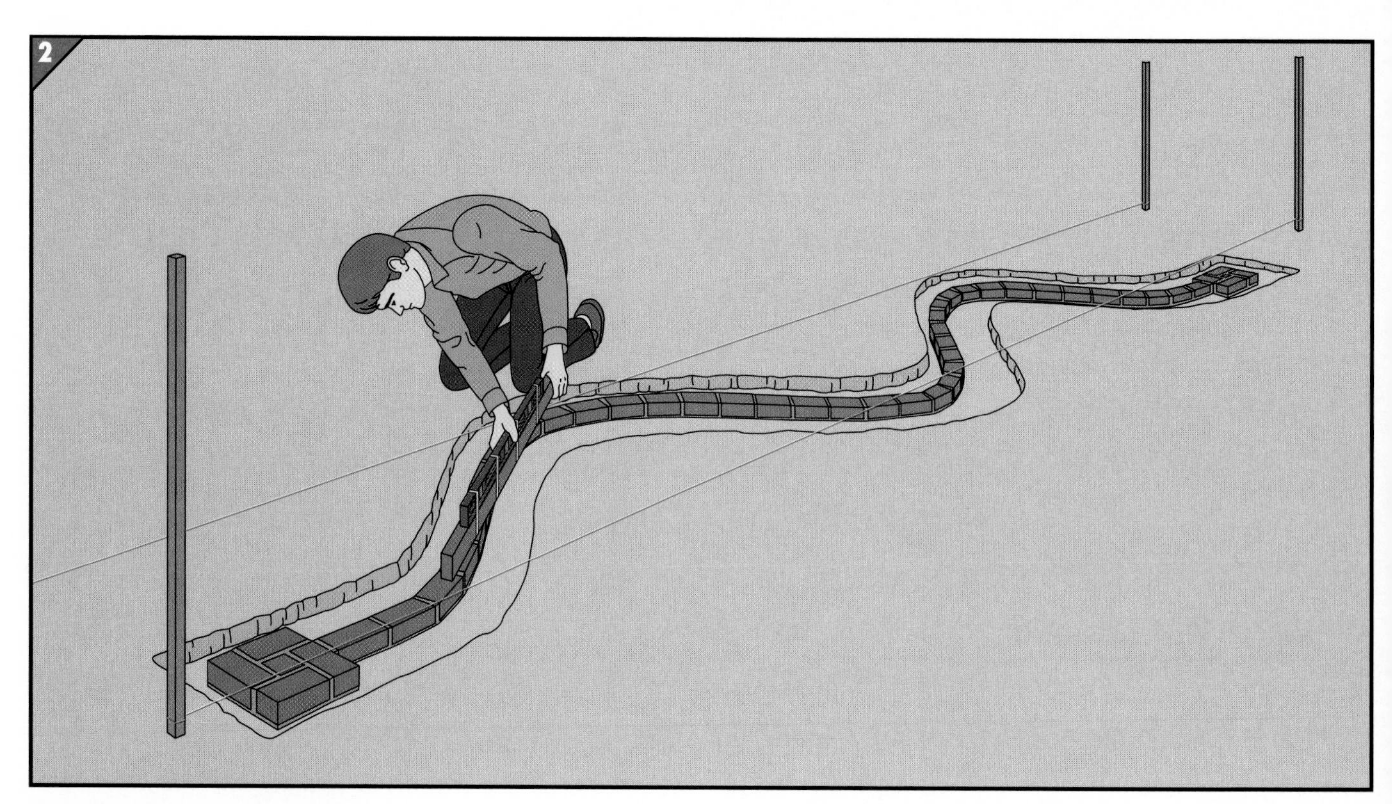


### 2. Checking for alignment.

◆ Mortar the first course of bricks for the pier.
◆ Remove three or four bricks at a time along the length of the dry run, then mortar them in place.
◆ Check that the tops of the outermost bricks along each curve align with the reference strings.
◆ Check the intermediate bricks for level by spanning a section of curve with a 2-by-4 topped by a level secured with tape. Position the board with one end resting on an outermost brick and pivot it *(above)*.
◆ To bring bricks into alignment, tap them with a trowel handle to lower them; or remove them and add more mortar to raise them.
◆ Lay the bricks for the other pier in mortar.

**3. Laying the second course.**

◆ Add a second course of bricks, moving the reference strings to the new course level. At the piers, arrange the bricks so their vertical joints are at the midpoint of the bricks in the course below and the first brick that extends into the wall crosses the midpoint of the brick below.

◆ Check for alignment *(right)* as in Step 1, and make any necessary adjustments.

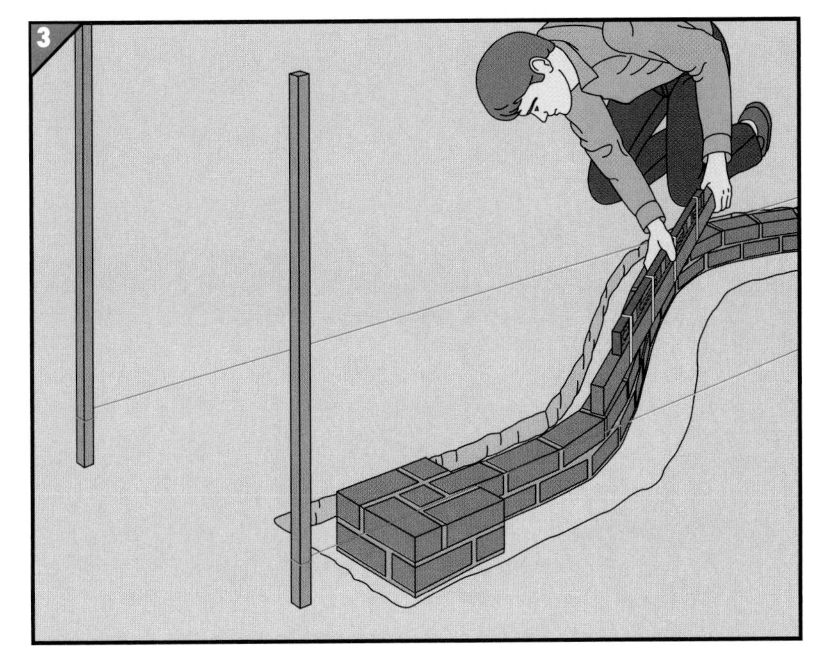

**4. Checking the curves.**

◆ After laying the second course, set the plywood template into the curves of the brickwork, supporting it with bricks at the level of the second course; look down onto the template to make sure that each end touches a brick at the outermost points of the curve.

◆ Bring the intermediate bricks into alignment by tapping them with the trowel handle until they lie flat against the template's curved edge *(left)*.

◆ Check along the length of the wall, flipping the template as necessary to match the curves.

◆ Check the other curves along the second course in the same way.

**5. Adding to the brickwork.**

◆ Raise the wall to the fifth course, resetting the string lines and checking each completed course for level, plumb, and correct curvature.

◆ Throw the mortar for the sixth course, then set a corrugated-steel wall tie across each vertical joint.

◆ Continue raising the wall *(right)*, setting ties at every fifth course. As the wall rises, you may need to stand on a stepladder and have a pair of helpers hold the template level against the course of bricks while you check the curvature.

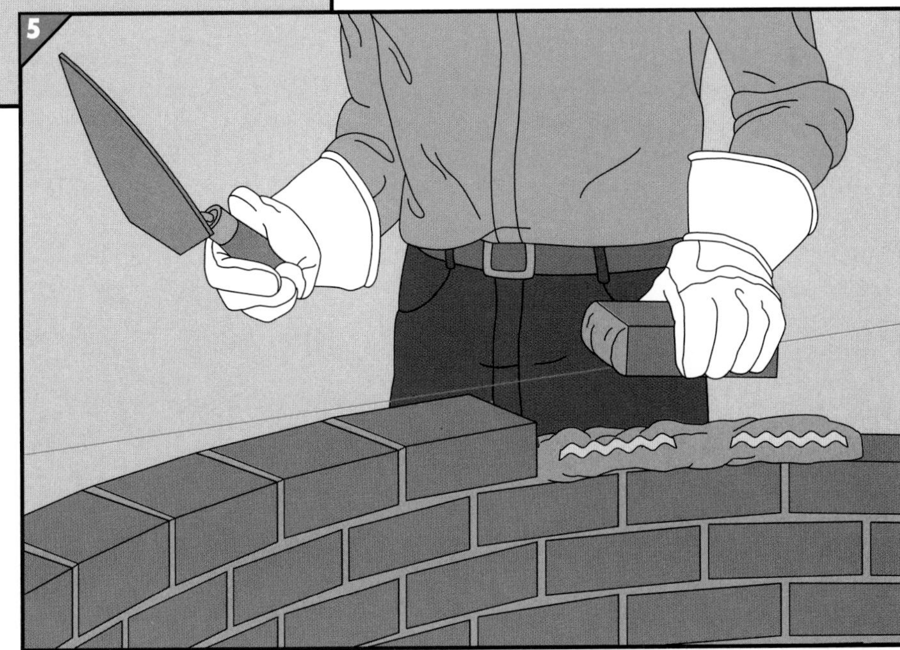

# Building Brick Arches

Designed to support the masonry above wall openings, arches are built for their beauty as much as for structural purposes. The two most common styles are the semicircular *(below)* and the jack, or flat *(opposite)*.

**Design and Construction:** For an arch to be self-supporting, its face must be at least as high as the thickness of the wall surrounding it, and its depth must equal that thickness. Although an arch can span more than 5 feet, such a project is best left to a professional. During construction, a semicircular arch is supported on temporary shoring referred to as a buck, which serves as a template while the bricks are being laid and carries the load until the mortar sets. Jack arches can be built to be self-supporting, but they are usually placed on permanent steel lintels. Buy a lintel that extends beyond the opening at least 8 inches on either side.

**Shaped Bricks:** If you build the arch with standard bricks, you will have to taper the mortar joints between them to fit. An alternative is to buy specially shaped bricks from a masonry supplier. Such bricks generally cost more, but they produce a better-looking arch. And in a jack arch, each brick is a slightly different size, so it is best to buy them ready-made. When ordering shaped bricks, supply the dealer with a full-size drawing of the arch, and plan the project well in advance since it will take some time for the shipment to arrive.

**TOOLS**

- Hammer
- Saber saw
- Mason's tape
- Level
- Mason's trowel
- Mason's line
- Jointer
- Circular saw with masonry blade
- Bricklayer's hammer
- Pointing chisel
- Maul
- Mason's hawk
- Joint filler

**MATERIALS**

- 2 x 4s
- Plywood ($\frac{1}{4}$", $\frac{1}{2}$")
- Common nails (3")
- Double-headed nails ($3\frac{1}{2}$")
- Masonry nails ($3\frac{1}{2}$")
- Roofing nails (1")
- Picture wire
- Bricks
- Mortar
- Jack-arch brick kit
- Steel lintel ($\frac{1}{4}$" x $3\frac{1}{2}$" x $3\frac{1}{2}$")
- Polyethylene sheeting (6-mil)
- Corrugated-steel wall ties

**SAFETY TIPS**

*Put on hard-toed shoes to protect your feet from falling bricks, and wear work gloves when handling bricks and working with mortar. Don goggles to drive nails or operate a power tool.*

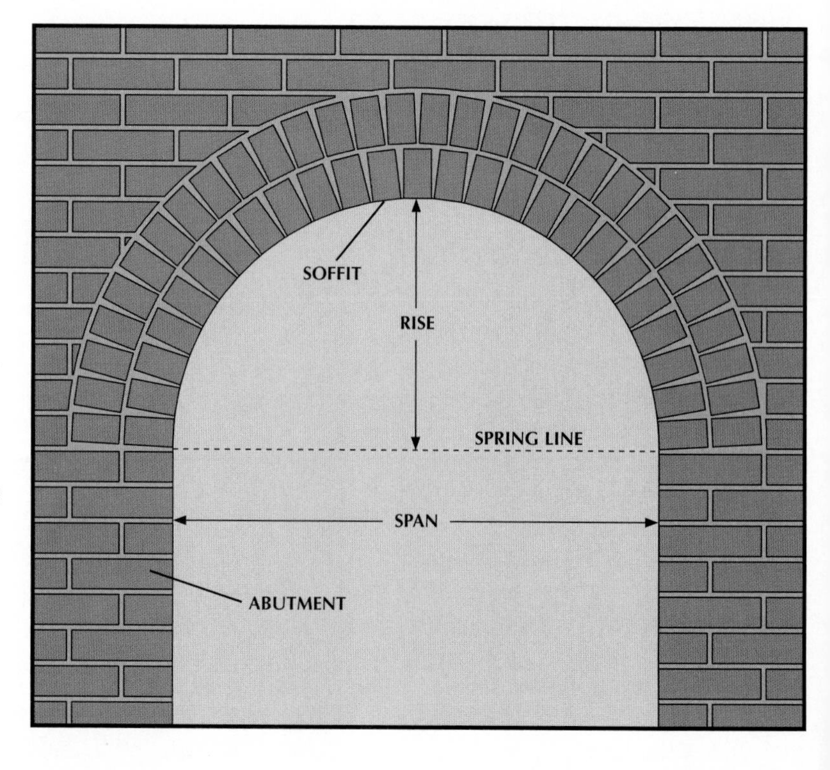


**A semicircular arch.**
This type of arch consists of two rings of standard-size bricks laid on edge. For the bricks to follow the curve of the arch, the mortar joints are made slightly wedge shaped—thicker at the top than at the bottom. Because the outside ring follows a wider radius than the inside ring, it requires a greater number of bricks.

The ends of the arch rest on masonry walls referred to as abutments; for this type of arch, the abutments are two bricks thick.

The imaginary horizontal line between the points where the arch meets the abutments is called the spring line. The distance between the abutments is the span, and the rise of the arch is the vertical distance from the spring line to the center of the underside of the arch. The entire underside of the arch is known as the soffit.

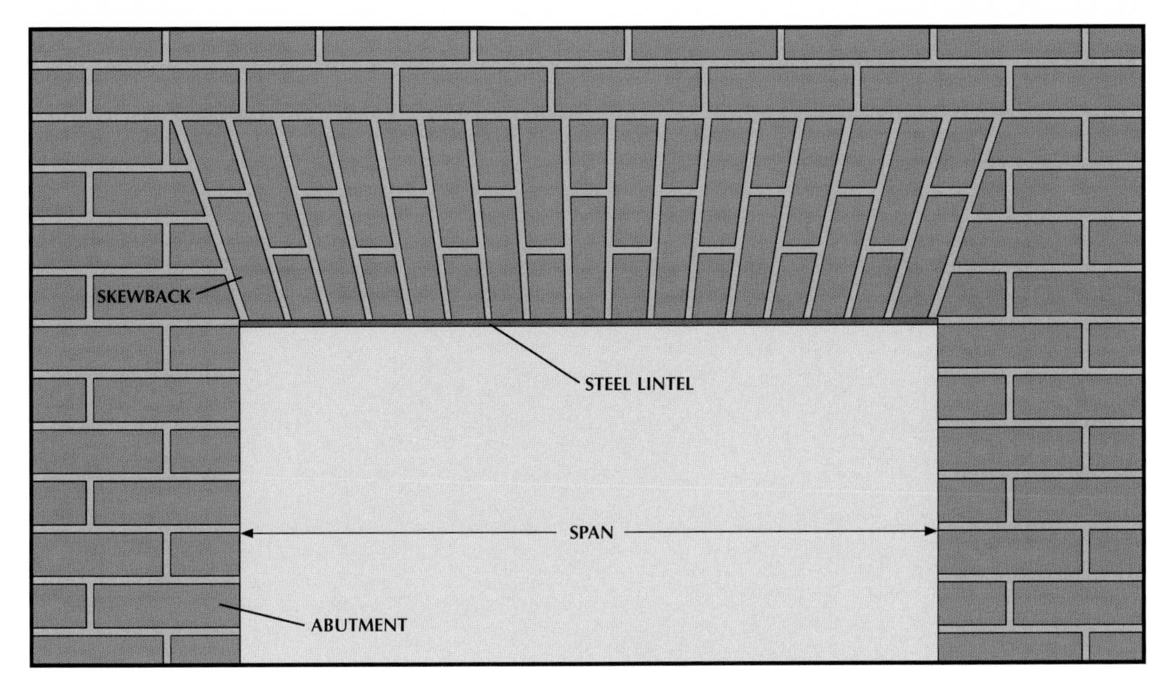


**A jack arch.**
Every brick in a jack, or flat, arch is cut to a slightly different shape, because each one sits at a slightly different angle to the steel lintel that supports it. The angle of the skewback—the inclined surfaces of the side walls, or abutments—is generally 70 degrees. When ordering a set of bricks for a jack arch, specify the span of the opening, the thickness of the mortar joints, and the height of the arch in terms of horizontal courses of surrounding brick (four, in this example). Other specifications are the depth of the arch and the pattern of the brickwork in the face of the arch—a running bond in this case.

## FORMING A SEMICIRCULAR OPENING

**1. Shaping the buck.**

◆ On a sheet of $\frac{1}{2}$-inch plywood, draw a spring line for the arch several inches in from the edge.

◆ Drive a nail at the center of the line and link a pencil to the nail with a piece of picture wire half the length of the spring line. Swing the pencil around to draw an arc connecting the ends of the spring line *(right)*.

◆ At each end of the spring line, continue the line straight to the edge of the plywood. Then, with a saber saw, cut along this line and around the arc.

◆ Using the cut plywood as a pattern, cut a matching piece.

◆ Nail a 2-by-4 spacer block on edge along the spring line between the two plywood pieces. Add shorter 2-by-4 spacers at intervals around the curved edge, fanning out from the center like the spokes of a wheel.

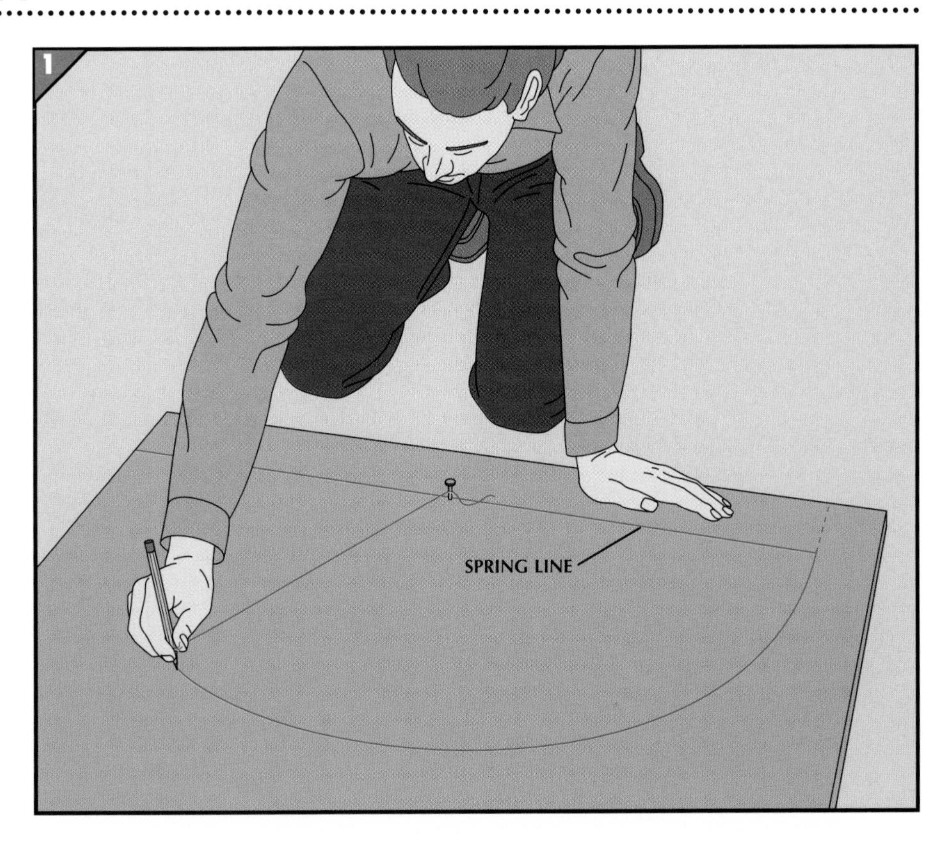

## 2. Marking brick positions.

◆ Lay the buck on the ground. Place one brick on end exactly at the center on top of the arch.
◆ Fill one side of the arch, spacing the bricks about $\frac{3}{8}$ inch apart and with the last brick $\frac{3}{8}$ inch above the spring line. Adjust the bricks if necessary, and check that the spacing is even by bending a mason's tape around the curve of the buck.
◆ Fill in the other side of the arch with the same number of bricks *(right)*.
◆ Mark all the brick positions on the face of the buck.
◆ Arrange a second ring of bricks around the first, starting $\frac{3}{8}$ inch above the spring line, leaving a $\frac{3}{4}$-inch gap between rings, and spacing the bricks evenly. Record the width at both ends of the wedge-shaped gaps between adjacent bricks.

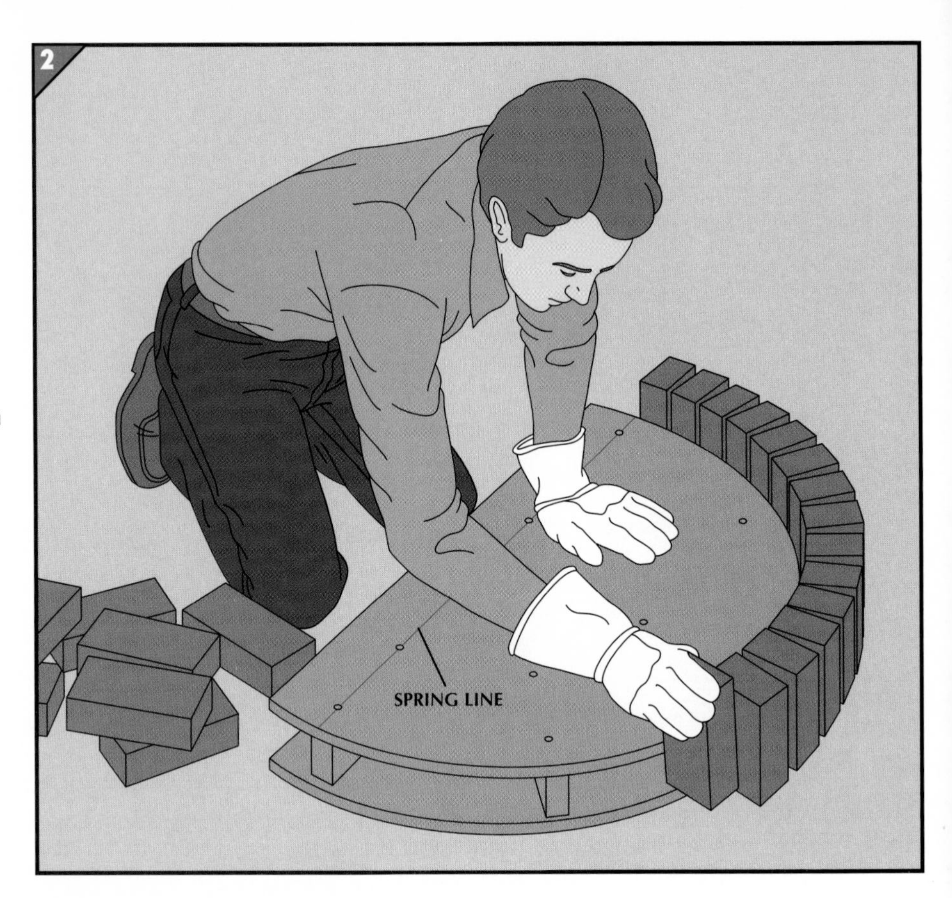


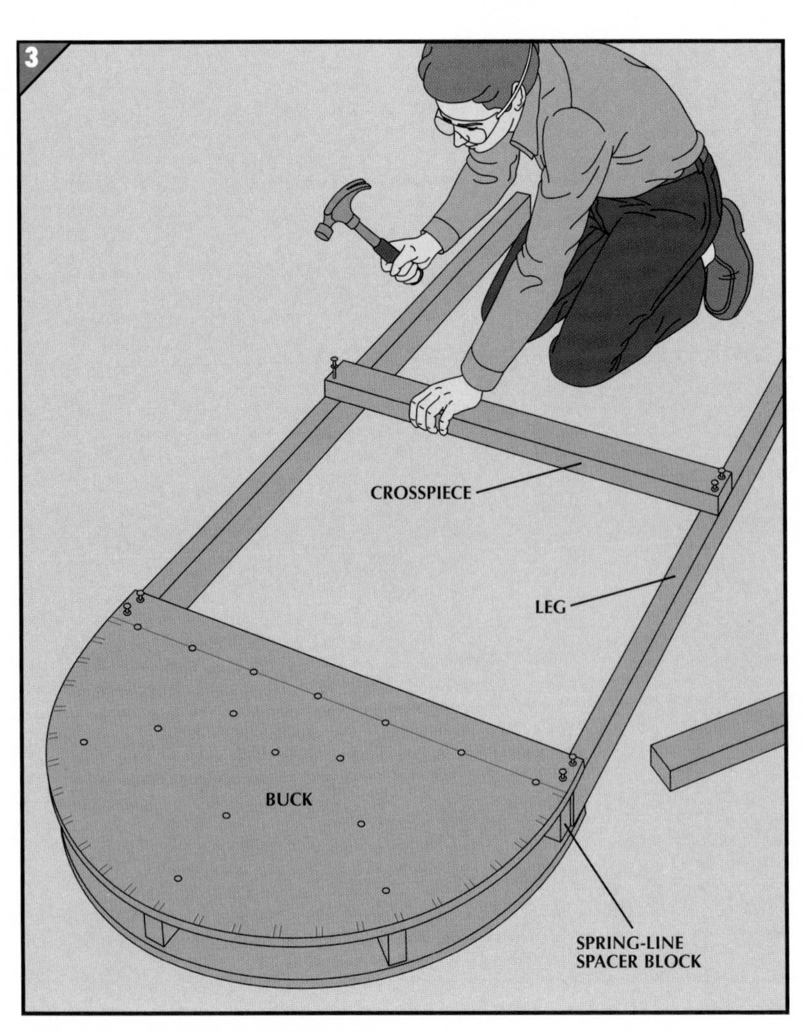


## 3. Setting up the buck.

◆ Hold the buck in place in the arch opening with the spring line level with the tops of the abutments, while a helper measures the distance between the spring-line spacer block and the ground. Cut two 2-by-4 legs to this length. Check the length of the legs by propping up the buck with each leg supporting one end of the spring-line spacer block.
◆ With $3\frac{1}{2}$-inch double-headed nails, fasten the legs in place between the plywood faces of the buck.
◆ Cut two 2-by-4 crosspieces equal in length to the span of the arch. Nail a crosspiece to one side of the pair of legs *(left)*. Turn the assembly over and nail a crosspiece to the other side of the legs.
◆ Raise the buck into the opening, hold a carpenter's level across the bottom edges of the two plywood pieces and, if necessary, slide shims beneath the legs to make the buck sit perfectly level.
◆ If the buck tends to tip, drive a masonry nail through each leg into mortar joints at the sides of the opening.

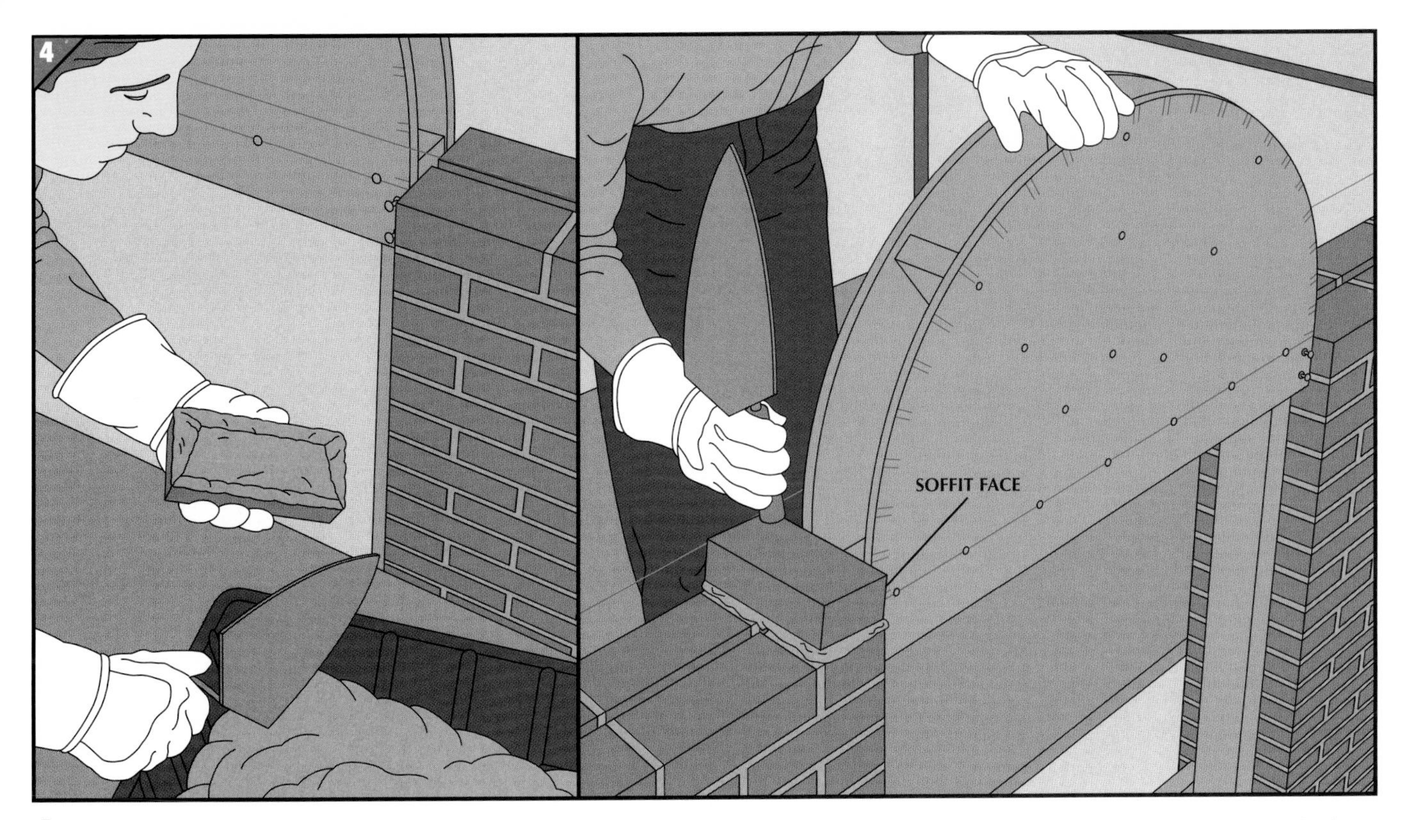


**4. Beginning the first ring.**

◆ Set up a mason's line level with the top of the first brick in the arch.

◆ Load a mason's trowel with mortar and shake the blade sharply downward. Then apply mortar along one bed of a brick, beveling the mortar so that there is a smaller amount on the part of the bed that will face in the direction of the opening *(above, left).*

◆ Lay the brick atop the corner of the abutment with its soffit face resting squarely against the buck. With the handle of the trowel, tamp the brick into the mortar *(above, right)* until it fits precisely between the marks on the buck, with its end flush with the mason's line.

◆ Continue laying bricks on both sides of the buck in this manner, raising the mason's line for each brick, until you reach the marks that you made on the buck to position the center brick.

TRICKS OF THE TRADE

## A Guide for Angled Bricks

The trick to creating a strong and even arch is to place each brick at the correct angle. This can be done by eye, but an easier and more accurate method is to use a guide. Cut a straight 1-by-2 slightly longer than the rise of the arch and nail one end at the center point of the spring line on the buck. As you place each brick, swing the stick around until it rests against the exposed edge of the brick. Adjust the angle of the brick until it matches the angle of the stick.

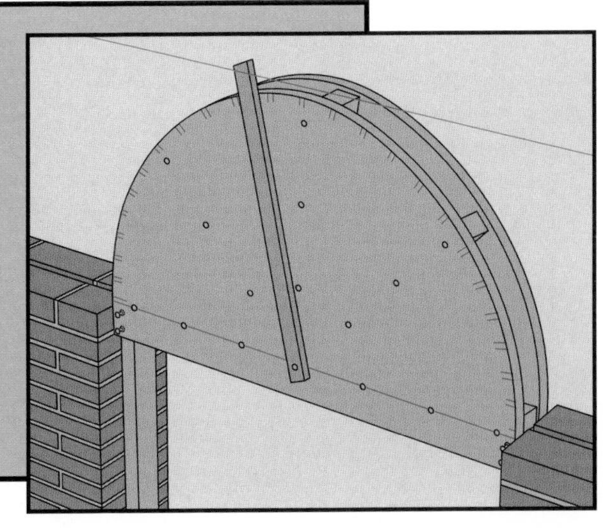

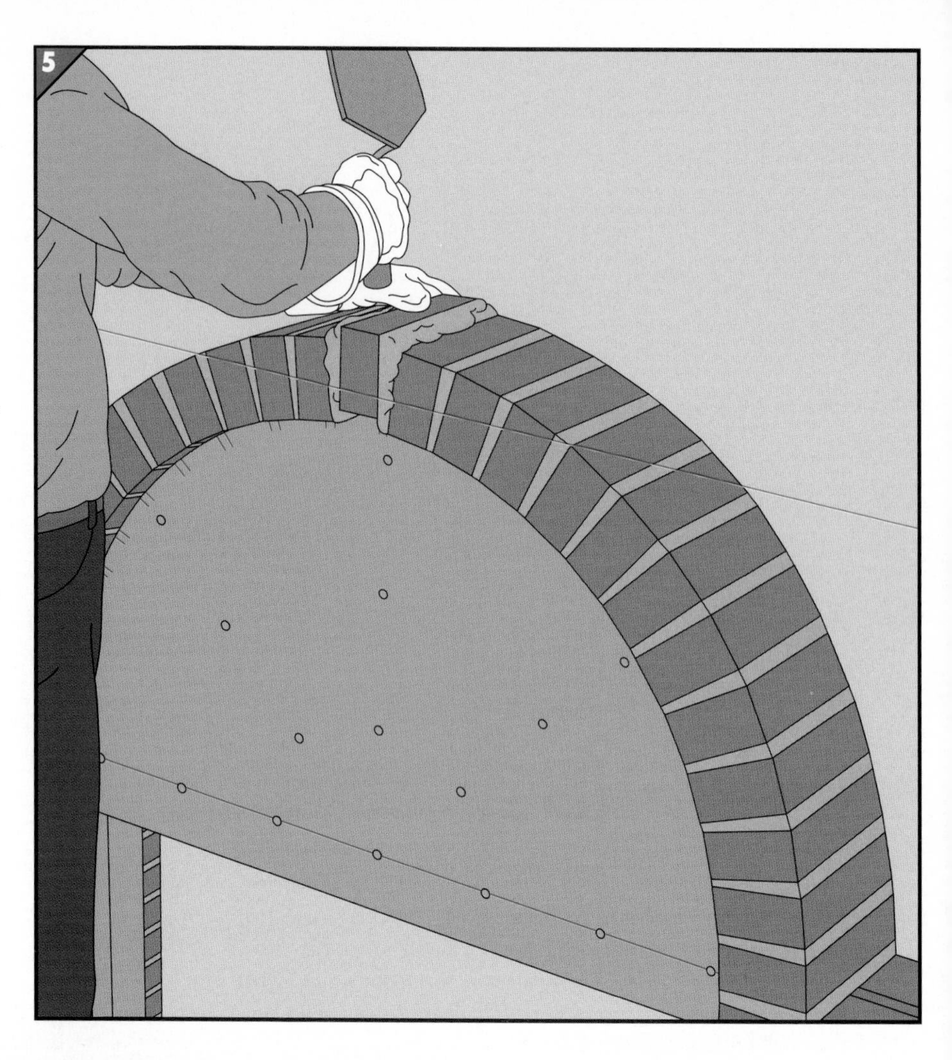


## 5. Laying the center brick.

◆ Butter the beds of both bricks adjoining the center brick.
◆ Butter the center brick on both beds, using the technique shown in Step 4.
◆ Slide the brick into place, tapping it with the trowel handle to wedge it against the buck *(right)*.

When the mortar has hardened enough to retain a thumbprint, tool all of the joints that will be visible on the completed arch as you would for the horizontal courses of the wall.

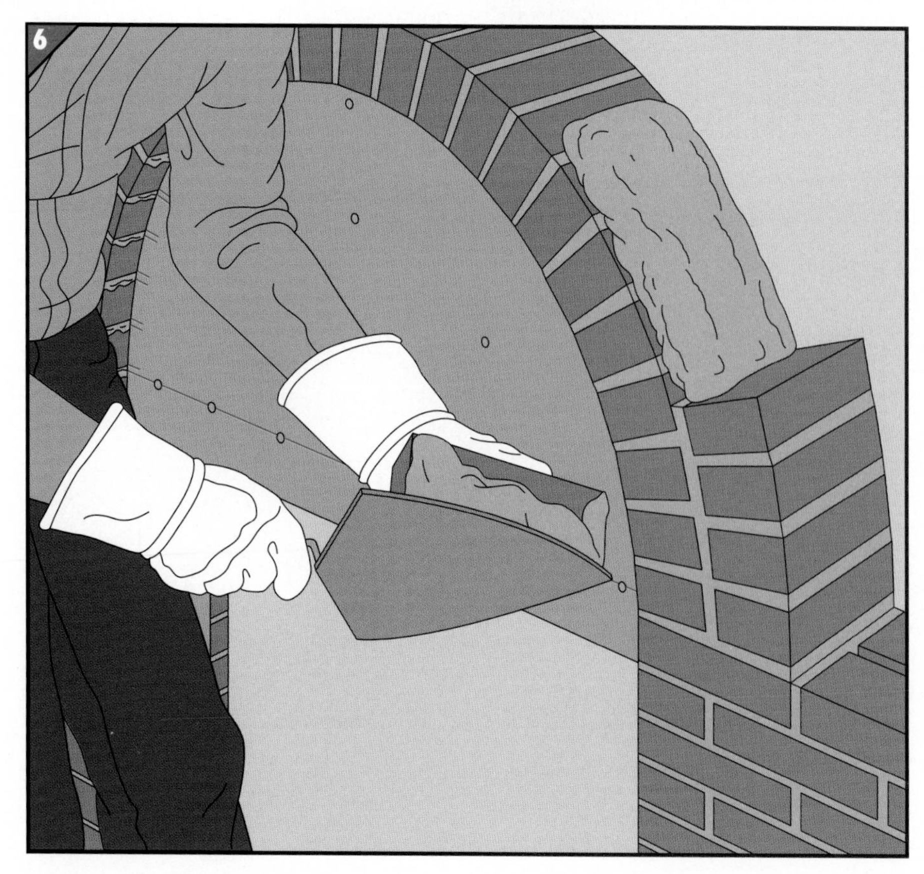


## 6. Building the outside ring.

◆ Butter one bed of a brick as illustrated in Step 4. Also spread mortar on the brick face that will lie against the inside ring.
◆ Lay the brick in place on top of one abutment, tapping it into the mortar until there is a $\frac{3}{8}$-inch joint between the end of this brick and the adjoining one in the first ring.
◆ Lay a few more bricks in the same way, creating wedge-shaped mortar joints following the measurements noted in Step 2.
◆ When you are partway up the curve of the arch, spread mortar along the top of the inside ring; continue laying bricks, buttering only one bed of each unit *(left)*.
◆ When you reach the middle of the arch, do not lay the center brick; build up the other side of the arch in the same way, and then lay the center brick *(Step 5)*.

**7. Continuing the wall upward.**

◆ Fill in horizontal courses of masonry on one side of the arch. When you get too close to the arch to fit a full brick, mark the outside face of a brick for a diagonal cut.

◆ Score the marked line $\frac{1}{2}$ inch deep with a circular saw fitted with a masonry blade; then finish the cut by chipping away the waste portion of the brick with the blade end of a bricklayer's hammer. Alternatively, make the cut with a masonry saw *(page 75)*.

◆ Butter the ends of the cut brick and slip it into place.

◆ Continue raising courses on both sides, cutting bricks as necessary to fit them against the arch *(left)*, until the arch is fully enclosed.

◆ Finish the joints.

**8. Finishing the arch.**

◆ After the mortar has cured for about five days, remove the pieces bracing the legs of the buck.

◆ With a helper supporting the buck, remove the nails fastening the legs and pry the bottoms of the 2-by-4s toward the center of the opening.

◆ Ease the buck out from under the arch.

◆ With a pointing chisel and a maul, cut out the hardened mortar from the joints of the soffit *(above)*, clearing them to a depth of $\frac{1}{2}$ inch.

◆ Load a hawk or the back of a trowel with mortar and hold it up close to the cleared joints. Force the mortar into the joints with a joint filler, packing them completely.

◆ When the mortar has set enough to hold a thumbprint, finish the soffit joints to match those on the face of the arch.

# SETTING A JACK ARCH

### 1. Making the pattern.

◆ From a piece of $\frac{1}{4}$-inch plywood, cut a full-size pattern of the planned arch.
◆ Taking care to arrange the arch bricks in the same configuration in which they were shipped, lay all the bricks in place on the pattern, leaving spaces for mortar joints between the bricks and at either end of the pattern *(right)*.
◆ Mark the positions of the mortar-joint gaps along the top and bottom edges of the pattern.
◆ Remove the bricks, again setting them down in exactly the order in which they were assembled.

### 2. Framing the arch.

◆ For the first course above the spring line, lay the end bricks no closer than 8 inches from the sides of the opening.
◆ Continue raising courses, offsetting each succeeding course one half brick back *(left)*, until you have reached the planned height of the arch on both sides of the opening.
◆ Install a steel lintel and polyethylene flashing *(page 84, Steps 1-2)*.
◆ Fill the space between the front edge of the lintel and the face of the bricks in the abutments by scraping a little mortar off a trowel onto the front edge of the lintel on both sides of the opening.

**3. Positioning the bricks.**

◆ Lay several bricks mortarless and $\frac{3}{8}$ inches apart on the lintel—enough bricks so that when the successive, stepped-back courses are laid, their ends will extend far enough over the top of the opening to be covered by the plywood pattern when held above the opening.
◆ Lay a mortar bed for the next course, but leave the vertical joints free of mortar, and step the course a half brick back from the center of the opening.
◆ Build each successive course in this way, until you reach a height equal to that of the plywood pattern *(left)*.
◆ Position bricks on the opposite side of the opening in the same way.

**4. Marking the skewback angle.**

◆ Align the bottom of the plywood pattern with the top edge of the lintel, and center the pattern in the opening.
◆ Follow the sides of the pattern with a pencil to mark the skewback bricks *(right)*; then mark the bricks on the other end of the pattern.
◆ Number the marked bricks from bottom to top, then remove all of the bricks that were installed in Step 3; scrape off any mortar adhering to their surfaces.
◆ Cut the marked bricks on the diagonal *(page 117, Step 7)*.

## 5. Building the skewbacks.

◆ With the plywood pattern resting on the lintel, lay the numbered cut bricks from Step 4 in their original positions but this time with mortar; tamp each brick into the mortar until it fits flush against the edge of the pattern *(right)*.

◆ Transfer the joint marks on the pattern to the backup wall behind the arch and to the front lip of the steel lintel with chalk.

◆ Stretch a mason's line even with the tops of the skewbacks.

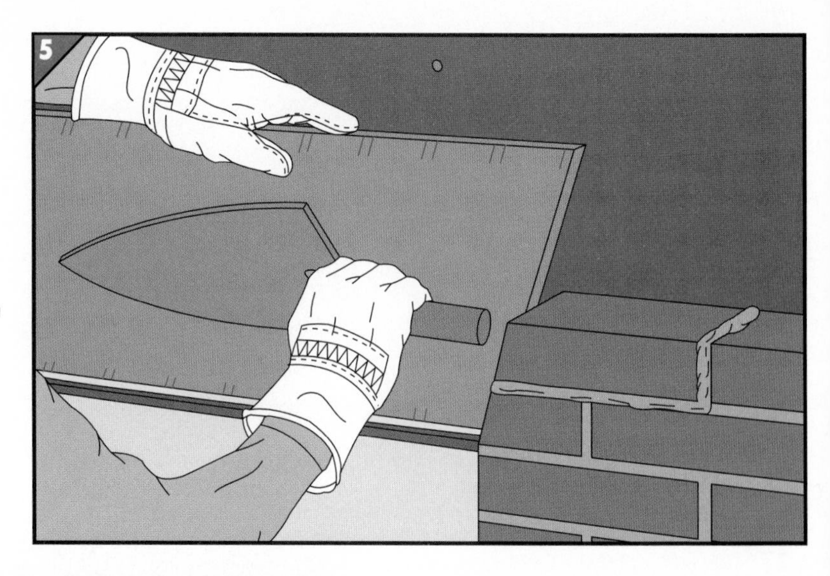

## 6. Beginning the arch.

◆ Butter one side of the bottom corner brick of the arch, with only a little mortar on the bottom part that will rest on the lintel *(left)*.

◆ Lay the brick in place against the skewback and tamp it into the mortar with the handle of the trowel until it is lined up with the marks on the lintel.

◆ Spread mortar on the top end of this first brick, then butter the half brick that fits on top of it; tamp the half brick in place, aligning its top with the mason's line and its sides with the marks on the backup wall.

◆ Install the first two bricks at the other end of the arch in the same way.

## 7. Completing the job.

◆ Working from both sides toward the center fill in all the remaining arch bricks—except the center ones—in their correct order, paying close attention to the thickness of the mortar joints *(right)*.

◆ Butter the center bricks on both sides and tap them into place with the handle of the trowel.

◆ Continue building the wall above the arch; to reinforce the top of the arch, lay a row of wall ties, end to end, in the mortar bed laid for the first course of bricks above the arch.

# Constructing Stone Arches

Whether spanning windows, doors, or fireplaces, or resting on a pair of piers to frame an entryway, stone arches lend a touch of timelessness to any setting.

**Planning Ahead:** The principles and procedures for building a stone arch are very similar to those for brick arches *(pages 112-120)*; however, the work is more exacting, and will probably require a helper on scaffolding on the other side of the wall as well to position the heavier stones. Since a stone arch is only as strong as its abutments and footing, the wall footing *(pages 44-50)* must continue along the opening, and the mortar in the abutments themselves—whether they are freestanding piers or walls *(pages 87-90)*—needs to set at least 24 hours before arch stones are laid. The piers for an arch that is not part of a larger wall—as well as the arch itself—need to be at least 12 inches thick.

A stone arch, like a stone wall, requires a stiff mortar mixture *(page 87)*, and the stones themselves must be strong, weather resistant, and workable—since you will need to do a lot of precise shaping *(pages 12-16)*. Single stones the thickness of the wall produce the strongest arch. However, smaller stones can also make up the required thickness, or be combined to make an arch slightly thicker than the wall, as a way of highlighting the arch. If you want to ensure precision, you can take your stone and a template of the arch to a professional stonemason to have the cuts made.

**Bracing the Arch:** The buck for a stone arch serves the same purpose as the one used for a brick arch *(pages 112-117)*. But because of the great weight of the stone, wooden shims are used to wedge the buck into place and, after the arch is built, they are pulled out to relieve the compression on the supports to allow the removal of the buck.

**TOOLS**

- Handsaw
- Saber saw
- Hammer
- Maul
- Stone chisel
- Stone hammer
- Pitching tool
- Bricklayer's hammer
- Mason's trowel
- Pointing trowel
- Stiff-bristle brush
- Level
- T-bevel
- Tuck pointer
- Pointing chisel
- Joint filler

**MATERIALS**

- 2 x 4s, 4 x 4s
- Plywood ($\frac{1}{2}$")
- Hardboard ($\frac{1}{4}$")
- Shims
- Double-headed nails ($3\frac{1}{2}$")
- Common nails (1", $3\frac{1}{2}$")
- Stones
- Mortar

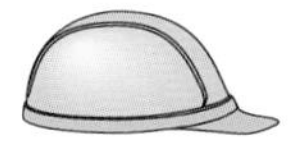

**SAFETY TIPS**

*Put on work gloves when working with mortar, and add hard-toed shoes when handling stone. Wear goggles to use power tools, to drive nails, and to cut stone.*

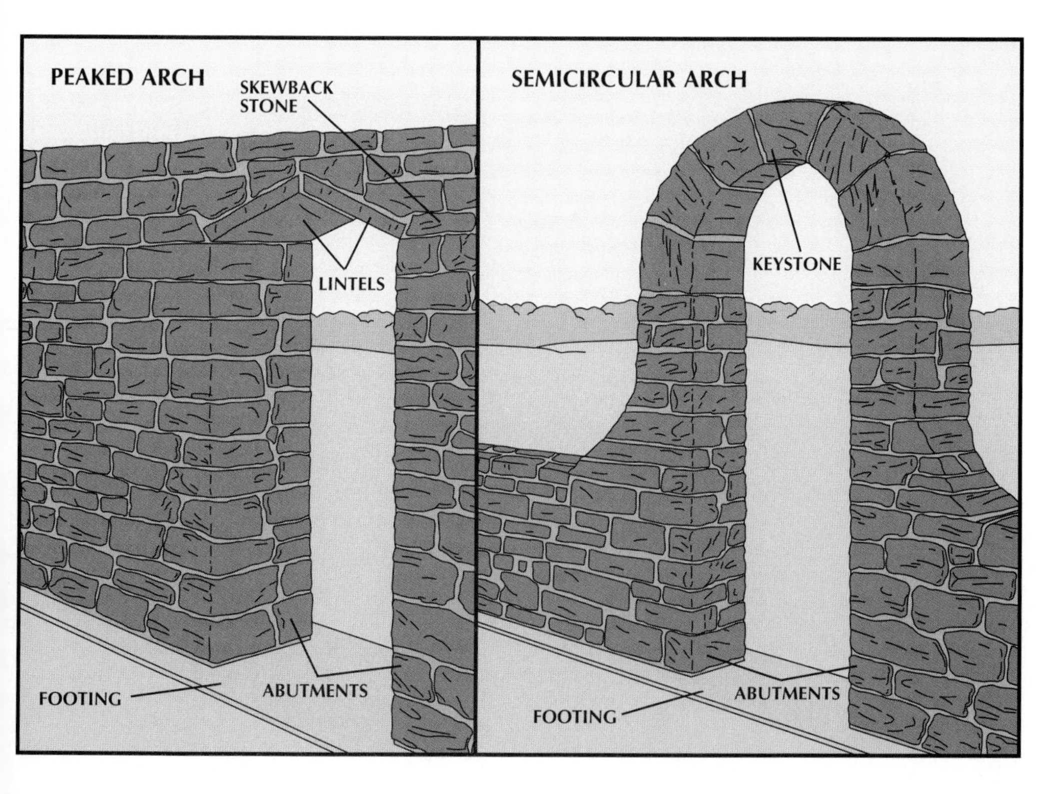


**Two types of arches.** A peak *(far left)* and a semicircle *(near left)* are two common designs for stone arches. The peaked arch, built over a triangular buck, has two paving stones that act as lintels and rest against skewback stones topping each abutment *(pages 122-123)*. The semicircular arch is made by setting large tapered stones over a semicircular buck; a keystone fits into position at the center *(pages 124-125)*.

# BUILDING A PEAK

## 1. Setting up the buck.

◆ Cut a triangle from $\frac{1}{2}$-inch plywood, with the base equal to the distance between the abutments and the distance to the peak equal to the desired rise from the base. Make another identical triangle then join the two with 2-by-4 spacers 2 inches shorter than the wall thickness.

◆ Trim four 4-by-4 posts $\frac{1}{2}$ inch shorter than the height of the spring line.

◆ For each pair of posts cut 2-by-4 spacers so that the assembly will be 1 inch narrower than the thickness of the wall, and toenail them to the posts at the top and bottom and every few feet in between with $3\frac{1}{2}$-inch common nails.

◆ To join the assemblies, cut 2-by-4s long enough to hold them firmly in place against the abutments, then toenail these spreaders to the top and bottom of each post.

◆ Position the buck on top of the post assembly, then tap in a wooden shim over each post *(right)*; continue tapping in one and then the other until the buck is at the spring line and level.

◆ With $3\frac{1}{2}$-inch double-headed nails, toenail up through the tops of the posts, through the shims, and into the buck.

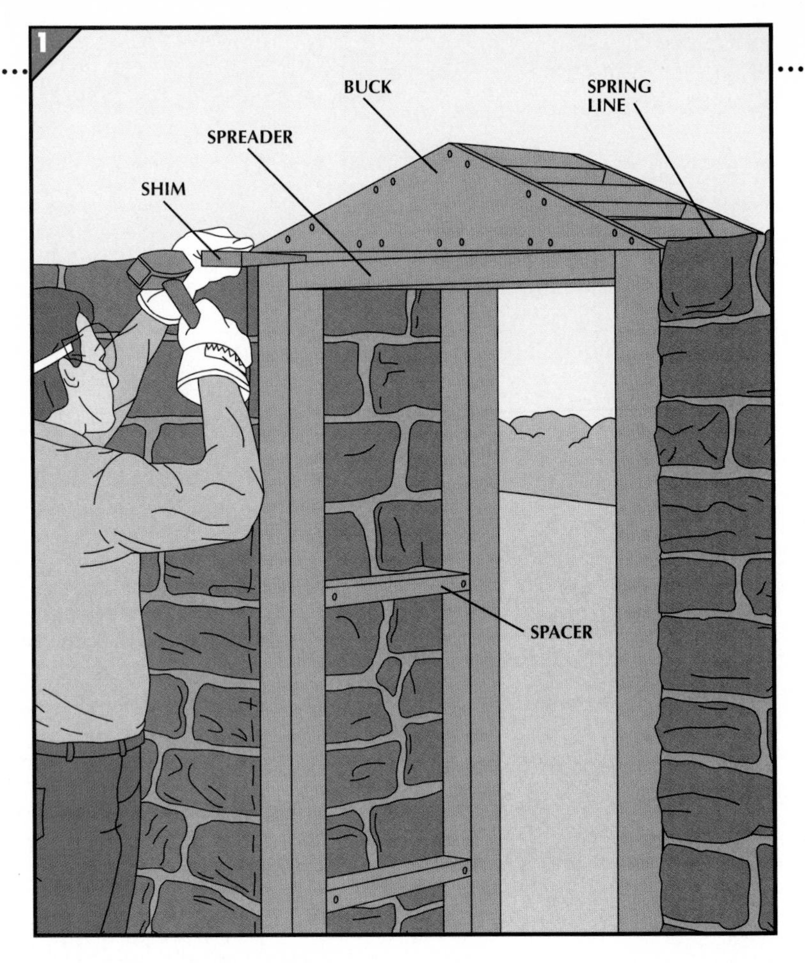

## 2. Setting the skewback stones.

◆ Select or cut stones that span the width of each abutment, trimming one side of each stone *(page 15)* to form an approximate right angle with the top surface of the buck.

◆ Lay the first skewback stone in a $\frac{3}{4}$-inch bed of mortar, with its angled end 1 inch away from the abutment edge *(left)*, then lay the opposite skewback stone in the same way.

◆ About an hour after the skewback stones have been laid, tool their wall-face joints *(page 90)*, then let the mortar set for at least 24 hours.

## 3. Mitering the lintel stones.

◆ Measure the distance from the peak of the buck to the bottom of the skewback stones, then select two paving stones for the lintels, each a few inches longer than this measurement and at least 4 inches thick.

◆ Subtract $\frac{3}{4}$ inch from the measurement to allow for mortar joints at each end and mark this length on one paving stone.

◆ Repeat this measuring and marking procedure for the paving stone on the other side of the arch.

◆ Holding a level perpendicular to the base of the buck, extend it past the buck's peak and set the arms of a T-bevel flush with the buck and the level *(right);* mark this angle on the stones.

◆ Cut the stones, taking care not to make them too short.

◆ Lay the lintels in place on top of the buck to check their fit, then remove and trim them if necessary.

## 4. Laying the lintels.

◆ Throw a $\frac{3}{4}$-inch bed of mortar on the angled side of one skewback stone.

◆ Dampen the first lintel with water and, with a helper, lower it into place *(left)*, sliding it down until the end rests in the mortar, and shifting it from side to side to center it; press it against the mortar until the joint is $\frac{1}{2}$ inch thick.

◆ Set the other lintel in the same way.

◆ Tamp mortar into the joint at the peak with a $\frac{1}{2}$-inch tuck pointer; allow the mortar to set for at least 24 hours.

◆ Build the wall in level courses over the arch, beveling the stones adjacent to the lintel for a close fit *(inset)*.

◆ Remove the shims and buck after 10 days, then repoint the joints on the underside of the lintel by first cutting and refilling them *(page 117, Step 8)* and then tooling them *(page 90)*.

# CREATING A HALF-ROUND PORTAL

### 1. Building the buck.

◆ Make a semicircular buck *(page 113, Step 1)* but using the edge of the plywood as the spring line and making it $\frac{1}{2}$ inch less than the span between the abutments; join the two pieces of plywood together with 2-by-4 spacers 2 inches shorter than the wall thickness.
◆ Cut a strip of $\frac{1}{4}$-inch hardboard as wide as the thickness of the buck.
◆ Nail the hardboard over the arch of the buck with 1-inch nails, starting at one end and bending the hardboard gently as you fasten it to the edges of the plywood *(right)*; at the opposite end, cut away any excess hardboard.
◆ Cut a marking string about 18 inches longer than the radius of the arch, lay the buck on its side, and attach the string to the nail used to draw the semicircle.

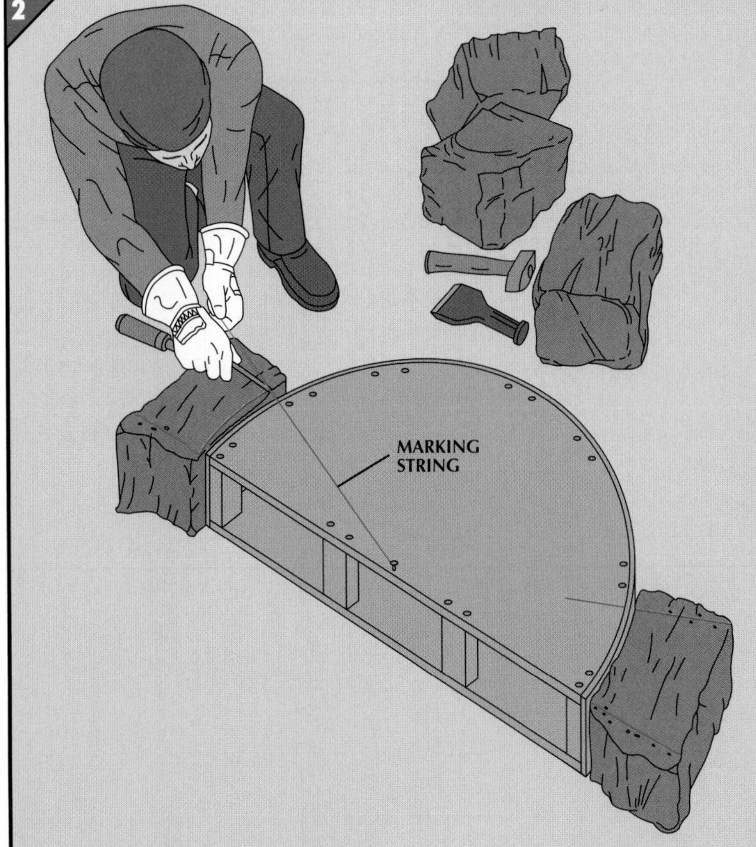


### 2. Positioning the stones.

◆ Select similarly sized stones as thick as the wall for the ends of the arch.
◆ Position a stone flush against the arch at each end of the buck. If it doesn't lie flat against the buck, scratch a line on it parallel to the edge of the buck with a tuck pointer.
◆ Pull the marking string over the the form's spring line and scratch a cutting line on the end of the stone.
◆ Mark the other edge of the stone in the same way—drawing a corresponding line on the buck —then mark both edges of the other stone *(left)*.
◆ Remove the stones and cut along the scribed lines *(page 15)*.
◆ Using the lines on the buck as references, leave $\frac{1}{2}$- to $\frac{3}{4}$-inch gaps for mortar joints and mark the bottom edges of a second pair of stones.
◆ Continue marking and cutting pairs of stones on opposite sides of the buck, working toward the top; mark their positions on the buck and allow for a single keystone at the top.
◆ Number the stones to indicate their positions for assembly.

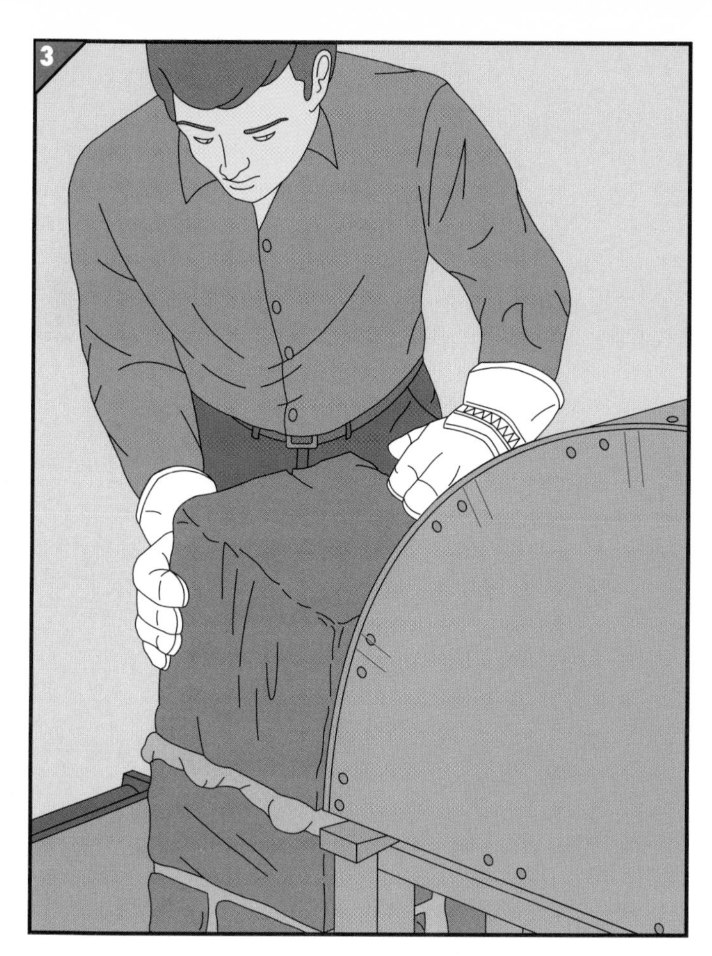

### 3. Laying the stones.

◆ Set up the buck on a post assembly in the opening *(page 122, Step 1),* then lay a $\frac{3}{4}$-inch bed of mortar on one abutment.
◆ Dampen the first stone slightly and center it on the mortar bed, twisting or rocking it lightly until it presses the mortar to a $\frac{1}{2}$-inch joint.
◆ Remove the excess mortar with a trowel.
◆ Set a stone on the opposite abutment in the same manner.
◆ Lay successive pairs of stones until only the gap for the keystone remains.

### 4. Setting the keystone.

◆ Butter both sides of the stones adjoining the keystone with mortar, then butter both sides of the keystone.
◆ Slide the keystone into the space at the top of the arch, centering it carefully.
◆ Finish all visible joints, striking them to match the depth of the joints in the abutments *(page 90).*
◆ Allow the mortar to cure for 10 days, then remove the buck.
◆ Repoint the underside of the arch, by first cutting a $\frac{1}{2}$-inch recess at each joint *(page 117, Step 8),* then finishing the joints to match the rest of the arch.

# INDEX

Time-Life Books is a division of Time Life Inc.

**TIME LIFE INC.**
PRESIDENT and CEO: George Artandi

**TIME-LIFE BOOKS**
PRESIDENT: Stephen R. Frary
PUBLISHER/MANAGING EDITOR: Neil Kagan

---

**HOME REPAIR AND IMPROVEMENT: Advanced Masonry**
*EDITOR:* Lee Hassig
*DIRECTORS OF MARKETING:* Steven Schwartz, Wells P. Spence
*Art Director:* Kate McConnell
*Associate Editor/Research and Writing:* Karen Sweet
*Editorial Assistant:* Patricia D. Whiteford

---

*Director of Finance:* Christopher Hearing
*Directors of Book Production:* Marjann Caldwell, Patricia Pascale
*Director of Operations:* Betsi McGrath
*Director of Photography and Research:* John Conrad Weiser
*Director of Editorial Administration:* Barbara Levitt
*Production Manager:* Marlene Zack
*Quality Assurance Manager:* James King
*Library:* Louise D. Forstall

**ST. REMY MULTIMEDIA INC.**
*President and Chief Executive Officer:* Fernand Lecoq
*President and Chief Operating Officer:* Pierre Léveillé
*Vice President, Finance:* Natalie Watanabe
*Managing Editor:* Carolyn Jackson
*Managing Art Director:* Diane Denoncourt
*Production Manager:* Michelle Turbide

Staff for *Advanced Masonry*

*Series Editors:* Marc Cassini, Heather Mills
*Art Director:* Michel Giguère
*Assistant Editor:* Liane Keightley
*Designers:* Jean-Guy Doiron, Robert Labelle
*Editorial Assistant:* George Zikos
*Coordinator:* Dominique Gagné
*Copy Editor:* Judy Yelon
*Indexer:* Linda Cardella Cournoyer
*Systems Coordinator:* Éric Beaulieu
*Technical Support:* Jean Sirois
*Other Staff:* Lorraine Doré, Geneviève Dubé, Anne-Marie Lemay, Jenny Meltzer, Rebecca Smollett

## PICTURE CREDITS

*Cover:* Photograph, Robert Chartier. Art, Maryo Proulx. Bricks provided by Briqueterie St-Laurent.

*Illustrators:* Jack Arthur, La Bande Créative, Gilles Beauchemin, Frederic F. Bigio, Edward L. Cooper, Roger C. Essley, Chuck Forsythe, Gerry Gallagher, Adsai Hemintranont, William J. Hennessy Jr., Elsie J. Hennig, Walter Hilmers Jr., Fred Holz, John Jones, Arezou Katoozian, Dick Lee, James Robert Long, John Martinez, John Massey, Peter McGinn, Joan S. McGurren, Bill McWilliams, Eduino Pereira, Jacques Perrault, Ray Skibinski, Snowden Associates Inc., Wagner/Graphic Design, Whitman Studio Inc.

*The following illustrations are based on material from:* **29, 31:** Lynn Ladder & Scaffolding Co., Inc. **38, 39, 40, 43:** Laser Tools Co.

*Photographers:* **End papers:** Glenn Moores and Chantal Lamarre. **12, 19:** Robert Chartier. **24:** DESA International. **25, 39:** Robert Chartier. **44:** London Machinery Inc. **47:** Robert Chartier. **51:** Vermeer Manufacturing Co. **55:** Wacker Corporation. **61:** Robert Chartier. **65:** American Honda Motor Co., Inc. **73:** Robert Chartier. **75:** Felker Corporation. **83:** Stanley Goldblatt Tools, Div of Stanley Works. **98:** Robert Chartier.

**ACKNOWLEDGMENTS**
The editors wish to thank the following individuals and institutions: Aearo Company, Southbridge, MA; American Honda Motor Co., Inc., Duluth, GA; Barrett Manufacturing Co., Chicago, IL; Bon Tool Co., Gibsonia, PA; Brick Institute of America, Reston, VA; Dayton Superior Canada Ltd., Montreal, Que.; DESA International, Bowling Green, KY; Larry Ford, City of Stockton, Stockton, CA; Louis Genuario, Genuario Construction Co., Inc., Alexandria, VA; Glen-Gery Corp., York, PA; Groupe Permacon Inc., Ville d'Anjou, Que.; Guy Guénette Ltd., St. Laurent, Que.; Hillman Fasteners, Cincinnati, OH; Alan David Kline, President, Lynn Ladder & Scaffolding Co., Inc., West Lynn, MA; Danielle Lane, Briqueterie St-Laurent, La Prairie, Que.; Laser Tools Co., Little Rock, Arkansas; London Machinery Inc., London, Ont.; The L.S. Starrett Company, Athol, MA; Marshalltown Trowel Company, Marshalltown, IA; Milwaukee Electric Tool Corp., Brookfield, WI; MKM Communications, Schaumburg, IL; National Concrete Masonry Association, Herndon, VA; Portland Cement Association, Skokie, IL; RCP Block and Brick Inc., Lemon Grove, CA; S-B Power Tool Co., Chicago, IL; Sears Craftsman, Hoffman Estates, IL; Service of the National Center for Earthquake Engineering Research, Buffalo, NY; Stanley Goldblatt Tools, Div. of Stanley Works, New Britain, CT; U.S. Ancho Corp., Pompano Beach, FL; Vermeer Manufacturing Co., Pella, IA; Wacker Corp., Menomonee Falls, WI; Webster & Fils Ltd., Montreal, Que.


First printing. Printed in U.S.A.
Published simultaneously in Canada.
School and library distribution by Time-Life Education, P.O. Box 85026, Richmond, Virginia 23285-5026.



**Library of Congress Cataloging-in-Publication Data**
Advanced masonry / by the editors of Time Life Books.
p. cm. — (Home repair and improvement)
Includes index.
ISBN 0-7835-3915-0
1. Masonry—Amateurs' manuals.
I. Time-Life Books. II. Series.
TH5313.A38 1998
693'.1—dc21 97-43435